Periodic table of the elements and element atomic weights (adapted from IUPAC 1991 values)

1 IA IA	2 IIA IIA	3 IIIA IIIB	4 IVA IVB	5 VA VB	6 VIA VIB	7 VIIA VIIB	8 VIIIA VIIIB	9 VIIIA VIIIB	10 VIIIA VIIIB	11 IB IB	12 IIB IIB	13 IIIB IIIA	14 IVB IVA	15 VB VA	16 VIB VIA	17 VIIB VIIA	18 VIIIB VIIIA
1 **H** 1.008																	2 **He** 4.003
3 **Li** 6.941	4 **Be** 9.012											5 **B** 10.811	6 **C** 12.011	7 **N** 14.007	8 **O** 15.999	9 **F** 18.998	10 **Ne** 20.180
11 **Na** 22.990	12 **Mg** 24.305											13 **Al** 26.982	14 **Si** 28.086	15 **P** 30.974	16 **S** 32.066	17 **Cl** 35.453	18 **Ar** 39.948
19 **K** 39.098	20 **Ca** 40.078	21 **Sc** 44.956	22 **Ti** 47.88	23 **V** 50.942	24 **Cr** 51.996	25 **Mn** 54.938	26 **Fe** 55.847	27 **Co** 58.933	28 **Ni** 58.693	29 **Cu** 63.546	30 **Zn** 65.39	31 **Ga** 69.723	32 **Ge** 72.61	33 **As** 74.922	34 **Se** 78.96	35 **Br** 79.904	36 **Kr** 83.80
37 **Rb** 85.468	38 **Sr** 87.62	39 **Y** 88.906	40 **Zr** 91.224	41 **Nb** 92.906	42 **Mo** 95.94	43 **Tc** (97.907)	44 **Ru** 101.07	45 **Rh** 102.906	46 **Pd** 106.42	47 **Ag** 107.868	48 **Cd** 112.411	49 **In** 114.818	50 **Sn** 118.710	51 **Sb** 121.757	52 **Te** 127.60	53 **I** 126.904	54 **Xe** 131.29
55 **Cs** 132.905	56 **Ba** 137.327	57–71	72 **Hf** 178.49	73 **Ta** 180.948	74 **W** 183.84	75 **Re** 186.207	76 **Os** 190.23	77 **Ir** 192.22	78 **Pt** 195.08	79 **Au** 196.967	80 **Hg** 200.59	81 **Tl** 204.383	82 **Pb** 207.2	83 **Bi** 208.980	84 **Po** (208.982)	85 **At** (209.987)	86 **Rn** (222.018)
87 **Fr** (223.020)	88 **Ra** 226.025	89–103	104 **Unq** (261.11)	105 **Unp** (262.114)	106 **Unh** (263.118)	107 **Uns** (262.12)	108 **Uno** (265)	109 **Une** (265)									

Lanthanides

57 **La** 138.906	58 **Ce** 140.115	59 **Pr** 140.908	60 **Nd** 144.24	61 **Pm** (144.913)	62 **Sm** 150.36	63 **Eu** 151.965	64 **Gd** 157.25	65 **Tb** 158.925	66 **Dy** 162.50	67 **Ho** 164.93	68 **Er** 167.26	69 **Tm** 168.934	70 **Yb** 173.04	71 **Lu** 174.967

Actinides

89 **Ac** 227.028	90 **Th** 232.038	91 **Pa** 231.036	92 **U** 238.029	93 **Np** 237.048	94 **Pu** (244.064)	95 **Am** (243.061)	96 **Cm** (247.070)	97 **Bk** (247.070)	98 **Cf** (251.080)	99 **Es** (252.083)	100 **Fm** (257.095)	101 **Md** (258.10)	102 **No** (259.101)	103 **Lr** (262.11)

Notes: Elements for which the atomic weight is contained within parentheses have no stable nuclides and the weight of the longest-lived known isotope is quoted. The three elements Th, Pa, and U do have characteristic terrestrial abundances and these are the values quoted. In cases where the atomic weight is known to better than three decimal places, the quoted values are rounded to three decimal places. The current IUPAC convention. The top, numeric, labelling system (1–18) is the current IUPAC convention. The other two systems are less desirable since they are confusing, but still in common usage. The designations A and B are completely arbitrary. The first of these (A left, B right) is based upon older IUPAC recommendations and frequently used in Europe. The last set (main group elements A, transition elements B) was in common use in America. For a discussion of these and other labelling systems see: Fernelius, W.C. and Powell, W.H. (1982). Confusion in the periodic table of the elements, *Journal of Chemical Education*, **59**, 504-508.

WORKBOOKS IN CHEMISTRY

SERIES EDITOR

STEPHEN K. SCOTT

WORKBOOKS IN CHEMISTRY

WORKBOOKS IN CHEMISTRY

Beginning Group Theory for Chemistry

PAUL H. WALTON

Department of Chemistry
University of York

OXFORD UNIVERSITY PRESS

This book has been printed digitally and produced in a standard specification
in order to ensure its continuing availability

OXFORD
UNIVERSITY PRESS

Great Clarendon Street, Oxford OX2 6DP

Oxford University Press is a department of the University of Oxford.
It furthers the University's objective of excellence in research, scholarship,
and education by publishing world-wide in

Oxford New York

Auckland Bangkok Buenos Aires Cape Town Chennai
Dar es Salaam Delhi Hong Kong Istanbul Karachi Kolkata
Kuala Lumpur Madrid Melbourne Mexico City Mumbai Nairobi
São Paulo Shanghai Taipei Tokyo Toronto

Oxford is a registered trade mark of Oxford University Press
in the UK and in certain other countries

Published in the United States
by Oxford University Press Inc., New York

ISBN 0-19-855964-X

Antony Rowe Ltd., Eastbourne

Workbooks in Chemistry: Series Preface

The new *Workbooks in Chemistry Series* is designed to provide support to students in their learning in areas that cannot be covered in great detail in formal courses. The format allows individual, self-paced study. Students can also work in groups guided by tutors. Teaching staff can monitor progress as the students complete the exercises in the text. The Workbooks aim to support the more traditional teaching methods such as lectures. The format of the Workbooks has been evolved through experience and discussions with students over several years. Students benefit through the Examples and Exercises that provide practice and build confidence. University staff faced with increasing class sizes may find Workbooks helpful in encouraging 'self-learning' and meeting the individual needs of their students more efficiently. The topics covered in the early Workbooks in the series will concentrate on background support appropriate to the early years of a Chemistry degree, including mathematics, performing calculations and basic concepts in organic chemistry. These should also be of interest to students who are taking chemistry courses as part of other degree schemes, such as biochemistry and environmental sciences. Later Workbooks will be designed to support material typically encountered in later years of a Chemistry course.

Preface

The writing of this book was prompted from two sources. First, the workbook format seems to be an ideal way of teaching group theory from a textbook; certainly, 'practice makes perfect' in chemical group theory. Second, group theory is one of those areas that chemists can find difficult—especially if they do not have a mathematical background. I hope that this book makes it clear that a knowledge of only the most basic arithmetic skills is required to use group theory successfully in chemistry. More importantly, I hope that students who might have been put off by the mathematics will discover that group theory is a powerful and beautiful theory in chemistry.

No book is produced by the efforts of a single person. This book is no exception, and I must thank all of those who have contributed to it. I thank my research group for being patient with me while I prepared the manuscript, and for being guinea pigs for some of the problems. Rajiv Bhalla and Spencer Harben checked many of the problems, and useful ideas were provided by colleagues Robin Perutz, Graham Doggett and Fred Manby. Of course, any mistakes in the book are my responsibility. Most of all I thank my family, in particular Daniel and Emma, for being understanding when I had to spend more than the usual amount of time at work completing the text.

York   Paul H Walton
September 1997

Contents

SECTION 1

What use is group theory?

What use is group theory?

There is no doubt that group theory is of pivotal importance in chemistry. In most undergraduate chemistry courses, the group theory course comes about halfway through the whole of the chemistry course. By this time the basic concepts used in chemistry will have been covered—especially the ideas of atomic orbitals, bonding theories, molecular vibrations, molecular shape and electronic configurations. To take these ideas further it becomes necessary to understand something about group theory. Indeed, courses which follow the group theory course are likely to build on the concepts (and terminology) of group theory quite extensively.

Unfortunately, it is true that some students can find their first encounter with group theory an uncomfortable one. In many cases this can be traced back to a lack of familiarity with the mathematical concepts that group theory uses, and in other cases it is related to the difficulty in 'seeing' the symmetry operations involved in examining a molecule's overall symmetry. Even for the students who are familiar with the mathematics and can perform three-dimensional manipulations in their head with little difficulty, the whole reason for group theory can be unclear.

Whatever your starting point when it comes to understanding group theory, it is important that you can at least use group theory to solve some chemical problems, and that you are familiar with the terminology, which pops up again and again in other areas of chemistry. Hopefully this book can help any chemistry student studying group theory for the first time. If you are one who finds group theory unpalatable because of the mathematics, then you should know that *using* group theory in chemistry only requires simple arithmetic: no more! (Of course, a fuller understanding of group theory requires mathematics.) In fact, the exercises in this book can be completed without having to read the 'mathematical' bit (Section 4) at all. If you are one of those who finds it difficult to 'see' the symmetry properties of a molecule, then realise that 'practice makes perfect'. The worked exercises in this book are designed exactly for this purpose, to make you more confident in handling the three-dimensional shape of molecules. And, finally, if you are one who just cannot see where group theory fits in and where it all leads, then some of the simple explanations in this book—particularly those in Section 5—will hopefully help to make things clearer. Certainly, doing the worked exercises will help you to see the use of group theory in chemistry.

This book is meant to be written in. Doing the exercises is an important part of completing the book, and you should have a pencil at hand. There are spaces for you to write your answers to the problems. The full answers are then given immediately below. Try to resist the temptation to look at the answer before you attempt the problem; it might help to cover·up the answer with a piece of paper. In saying this, do not expect to get every answer completely right first time, it would be remarkable if you did. It will also be very useful to have a small molecular modelling kit nearby (some 'Blu-Tack' and straws are a useful substitute if you do not possess a molecular modelling kit). Of course, some sections will take longer to complete than others, but it is probably realistic to try and complete each section in one sitting. Try to understand the ideas covered in each section before proceeding to the next one, and watch out for the terminology, which will keep popping up. By the end you will have covered nearly all of the uses of group theory in chemistry, to a first level of understanding. This should be enough at least to use group theory to solve problems that you are likely to encounter in chemistry, but should also provide a firm foundation if you are interested in taking group theory further.

SECTION 2

Symmetry—a start

Symmetry—a start

2.1 An introduction to symmetry

From an early age we have an understanding of symmetry. It is easy to spot a symmetrical object over an unsymmetrical object. These qualitative ideas also extend into chemistry. For instance, it is easy to say which of the molecules below is the most 'symmetrical'.

The most 'symmetrical' molecule is benzene. Why is it important to know this in chemistry? To answer this, one has only to look at the ^{1}H-NMR spectra of the two molecules. Despite the close structural similarity of the two molecules, the spectrum for benzene contains a *single* peak, whereas the spectrum for 2-chlorophenol contains several peaks split by a complicated coupling pattern. Clearly then, even at a qualitative level, symmetry has something to do with the energy levels of a molecule. We can make some 'hand-waving' arguments here that all of the benzene hydrogen atoms are in the same chemical environment, but it is difficult to go much further, and it is impossible to say anything more quantitative about how symmetry is important in determining the energy levels of a molecule. Therefore, if we are ever to say anything quantitative about the symmetry of a molecule—and clearly, this may prove to be extremely powerful in predicting energy levels and spectra of molecules—we must provide some type of rules to define the symmetry of a molecule. Indeed, this is the subject of this section: putting what we already do well qualitatively in determining the 'symmetry' of a molecule on to a more quantitative basis. The symmetry rules we shall consider can be applied to any finite object, but we shall only consider the rules in the context of the shape of molecules.

2.2 Rotation axes

The first type of symmetry which we will examine is symmetrical behaviour upon rotation. For example, look at the structure of the water molecule in the figure on the next page. The molecule has the well-known 'V-shape', and for our purposes imagine that the molecule is contained within the plane of the paper. We can imagine an axis, which passes through the oxygen atom and bisects the H–O–H angle, all in the plane of the paper. If the molecule is rotated by 180° about this axis, then we end up with the molecule which is *indistinguishable* from the original. In other words, the water molecule is symmetric to rotation by 180° about this axis. (The molecule after rotation is not *identical* to the molecule before rotation, but it is *indistinguishable*. If the hydrogen atoms are given imaginary labels, then H_1 and H_2 are interchanged upon rotation. Of course, the hydrogen atoms are the same in the real molecule.) Such symmetrical behaviour about an axis can be defined more mathematically as follows. If the molecule is symmetric by rotation of x degrees about an axis, then we have a rotation axis of order $(360/x)$. So, for water, the order of the axis is $(360/180) = 2$. Therefore, this is a second order axis, which is given the shorthand name C_2.

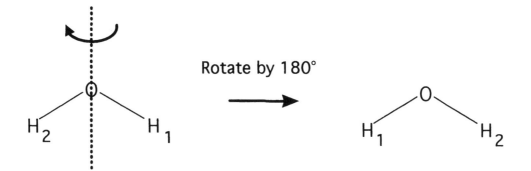

Rotate by 180°

Some notes of terminology.

- The axis is known as the *symmetry element*.

- The actual transformation of the molecule to its symmetrical equivalent position is known as the *symmetry operation*.

- The rotation is clockwise.

So, for the water molecule, the C_2 axis is the symmetry element, whereas the rotation of 180° about this axis is the symmetry operation.

Now, sketch what happens to the water molecule (labelling the hydrogen atoms as above), when the C_2 operation is carried out on it twice in succession.

$H_1 \quad O \quad H_2 \rightarrow H_2 \quad O \quad H_1 \rightarrow H_1 \quad O \quad H_2$

Answer

$$H_2 \quad O \quad H_1 \xrightarrow{\ C_2\ } H_1 \quad O \quad H_2$$

$$\xrightarrow{\ C_2\ } H_2 \quad O \quad H_1$$

The successive application of two C_2 operations on the water molecule, gives back the original molecule, which is *identical* to the molecule that we started with. Overall, nothing has happened to the molecule. In fact, the operation of doing nothing is an important one in group theory. As the example above shows, doing nothing can be seen as the multiple application of symmetry operations or as the rotation of the molecule by 360° about an axis (a C_1 axis) and therefore 'doing nothing' is a symmetry operation in its own right. Such a symmetry operation is denoted by the letter E. The importance of having a 'do nothing' operation will be seen in the next section.

The rotation operation by 180° about a C_2 axis is given the symbol $C_2{}^1$, and the overall operation of $C_2{}^1$ followed by another $C_2{}^1$ is given the symbol $C_2{}^2$. It can also be seen that $C_2{}^2$ is equivalent to E. In fact, the general case of $C_n{}^n$ (where n is a positive integer) is equivalent to E. This can be expressed more mathematically as follows:

$$C_2{}^1 . C_2{}^1 = C_2{}^2 = E$$

or, more generally $C_n{}^n = E$

Sketch the ammonia molecule and work out if it has a rotation axis (axes) and the order of the axis (axes). Draw the axis (axes) on your sketch of the molecule.

The molecule possesses a single rotation axis of order 3. The axis passes through the nitrogen atom and through the centre of an imaginary equilateral triangle formed by the three hydrogen atoms.

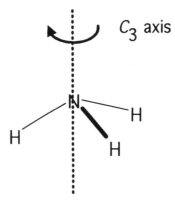

Clearly, the ammonia molecule possesses only one rotation axis of symmetry. What about molecules which possess more than one rotational axis of symmetry?

On the top of the next page sketch the structure of benzene. Figure out all of the rotation axes that can be found for benzene (a molecular model kit may be useful here), and sketch the axes on your picture of benzene.

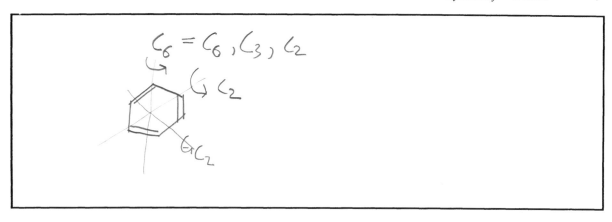

Answer

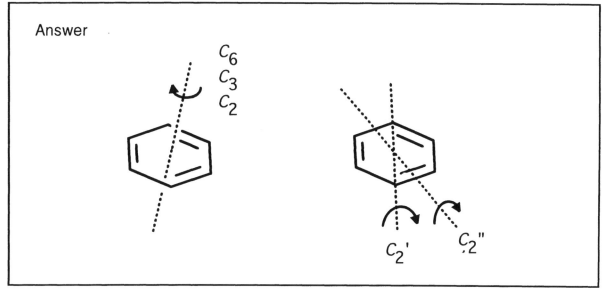

Benzene contains a surprisingly large number of rotation axes of symmetry, which are shown in the figure above. Note in particular that there are six C_2 rotation axes (only two are drawn in the figure) *perpendicular* to the highest order axis, which is a C_6 axis.

In molecules which have more than one rotational axis of symmetry, then the axis of highest order is defined as the *principal axis*. Therefore, in benzene, the principal axis is the C_6 axis. Also, the C_6 axis contains, by definition, a C_3 axis and a C_2 axis, which are co-axial with the C_6 axis. In other words, the following is true about symmetry operations:

$$C_6{}^2 = C_3{}^1$$
$$C_6{}^3 = C_2{}^1$$

In general, any even-order axis greater than order 2, will contain, lower-order even axes. For instance, a C_4 axis must contain a co-axial C_2 axis. A C_8 axis must contain both C_4 and C_2 axes.

Furthermore, in benzene, there are more C_2 rotation axes *perpendicular* to the principal axis. There are six such axes; three of which pass directly through atoms, the other three pass through the centres of C–C bonds (you should try and draw all of these axes on your sketch). It is necessary to distinguish these types of axes from the C_2 axis, which is coaxial with the principal axis. Thus, the two types of C_2 axes are called C_2' and C_2''.

2.3   Reflection planes

We have seen how part of a molecule's symmetry can be described by rotation axes. Are there other symmetry elements which we need to consider before we can fully describe the symmetry of a molecule in terms of symmetry elements? Well, of course, the answer is yes. The next most obvious symmetry operation is symmetrical behaviour upon reflection.

Again, we can start with the water molecule. Imagine a double-sided mirror plane which bisects the molecule, as shown in the figure on the top of page 8. The *symmetry operation* of such a mirror is to interchange the two hydrogen atoms (assuming that the mirror plane is exactly perpendicular to the plane of the molecule, and that it bisects the angle). The resulting molecule after the operation of the mirror is *indistinguishable* from the original in terms of shape—exactly analogous to the symmetry operation of the C_2 rotation axis. Therefore, this reflection plane is another *symmetry element* of the water molecule.

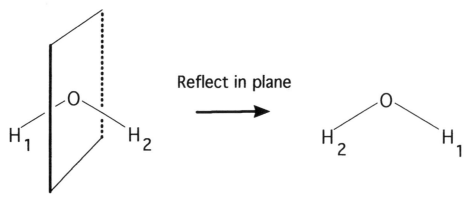

Sketch another reflection plane in the H$_2$O molecule.

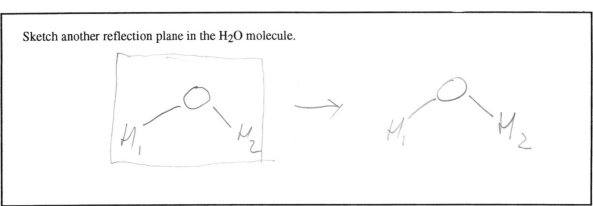

Answer

The other reflection plane is shown below. (Note that the water molecule has been drawn slightly out of the plane of the paper to give it more perspective.)

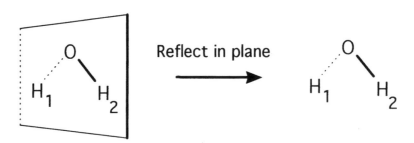

This reflection plane is within the plane of the molecule. Upon the symmetry operation defined by the plane the molecule is unchanged in shape, and it is also *identical* to the original molecule. The hydrogen atoms are not interchanged by this symmetry operation, and the starting and finishing molecules are *indistinguishable* in terms of shape, therefore this reflection plane is a symmetry element.

We can see that the water molecule contains two reflection planes, each perpendicular to the other. In group theory a reflection plane is denoted by the Greek letter sigma, σ. These reflection planes contain the principal axis within the plane, and can be said to be 'vertical' along with the principal axis. Accordingly, if a reflection plane contains a principal axis, it is denoted by the symbol σ$_v$, where '*v*' means 'vertical'.

List all of the symmetry elements present in the water molecule. Sketch the elements on a diagram of a water molecule.

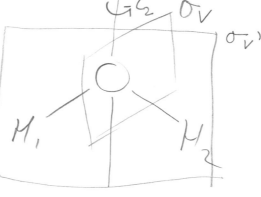

Answer

These are E, C_2, $\sigma_v(xz)$ and $\sigma_v'(yz)$. Notice that, in this case, the two σ_v planes are distinguished from each other by which cartesian plane they lie in. The xz plane is perpendicular to the plane of the molecule, with the z-axis pointing along the principal rotation axis.

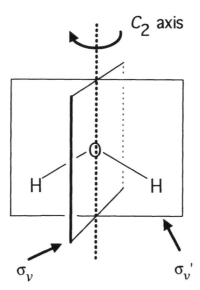

As above, identify the reflection planes in the NH_3 molecule. Sketch one of the reflection planes on a drawing of the molecule. Label the atoms and show the effect of the reflection plane on the position of the atoms.

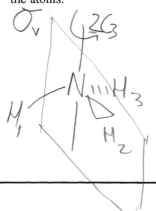

Answer

The NH_3 molecule is shown below with one of its three reflection planes. The NH_3 molecule is viewed 'from above' with the nitrogen atom nearest to the viewer. There are three reflection planes in total, each at 120° to each other. Each plane passes through a single N–H bond and exactly bisects the H–N–H angle opposite to the N–H bond. In the figure below, the H_b–N–H_c angle is bisected. The nitrogen and H_a atoms are *in the reflection plane itself*, whereas the other two hydrogen atoms lie either side of the plane. The operation of this reflection plane is to interchange the H_b and H_c atoms. The other two reflection planes will interchange different combinations of hydrogen atoms.

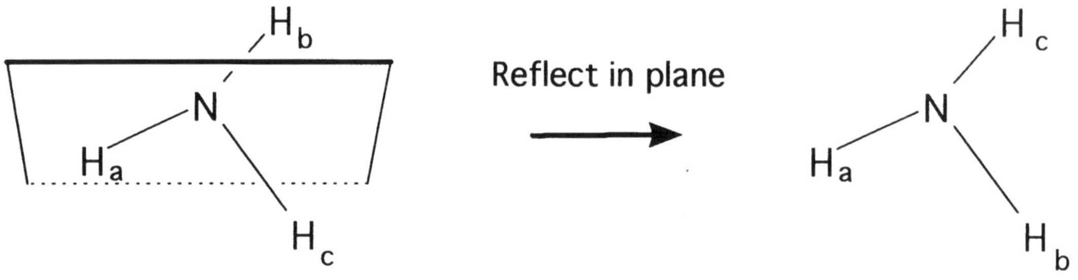

For the NH_3 example above, all three of the reflection planes are given the general symbol σ_v. This is because all of the planes contain the main C_3 rotation axis. Each plane can be uniquely identified by σ_v, σ_v' and σ_v''. This is analogous to the reflection planes in the water molecule.

There are two more types of reflection plane that we must consider. This is best done by looking at an example. Consider the planar molecule $AuBr_4^-$, which has a square planar structure around the gold atom.

Sketch the structure of $AuBr_4^-$, and identify the principal rotation axis in the molecule. Sketch this principal axis on your sketch. Identify any other rotation axes.

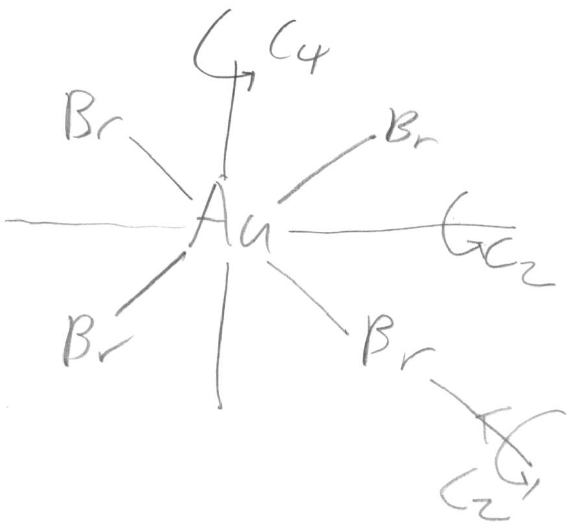

Answer

The molecule is sketched below with its principal C_4 axis. It is a C_4 axis, as the molecule can be rotated by 90° (360/4) around this axis, and be *indistinguishable* from the original molecule. Note also that there are C_2 axes perpendicular to the principal axis (analogous to the benzene molecule that we encountered above).

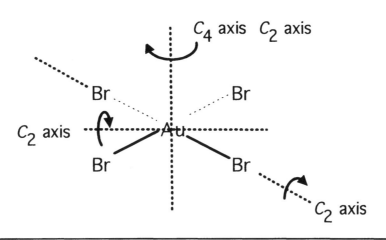

How many reflection planes can be found in this molecule? Perhaps the easiest to 'see' are the reflection planes which contain the principal axis. There are two types of such reflection planes. One type, which is marked as σ_v on the diagram below, contains the principal axis, and also contains Au–Br bonds. Note that there are actually two σ_v planes at 90° to each other (only one is drawn).

The other type of reflection plane is very similar to the σ_v planes insofar as they contain the principal axis, but they *do not* contain any of the Au–Br bonds; these planes *bisect* the Br–Au–Br angles. There are two of this type of plane, which are given the symbol σ_d, 'd' stands for 'dihedral'. Notice the similarity of the σ_v and σ_d planes, which are only different because the σ_v planes contain the Au–Br bonds, whereas the σ_d planes bisect the Br–Au–Br angles.

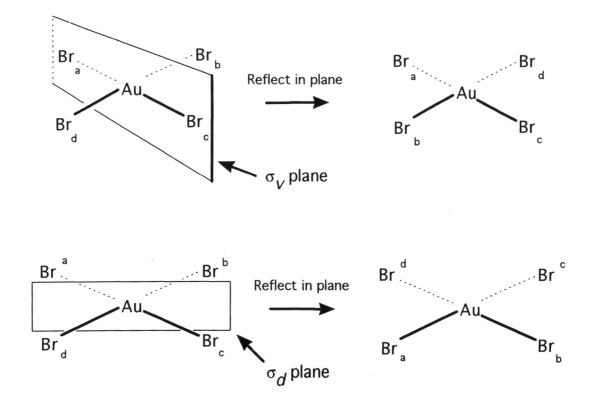

Sketch the AuBr$_4^-$ molecule and the σ_v and σ_d planes which are not shown in the diagram above.

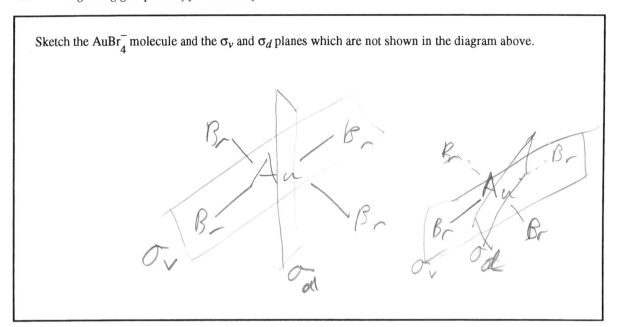

Answer

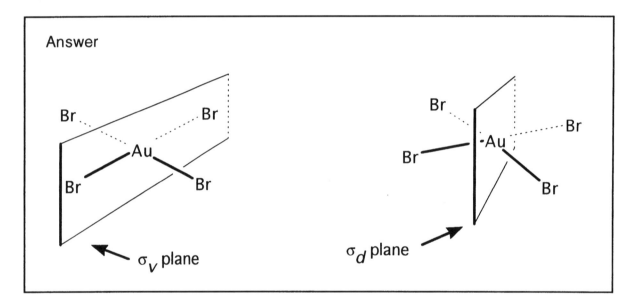

Apart from the σ_v and σ_d reflection planes shown above, there is one more reflection plane in the AuBr$_4^-$ molecule. Figure out where the new plane lies, and sketch the plane on a diagram of the molecule.

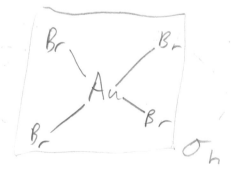

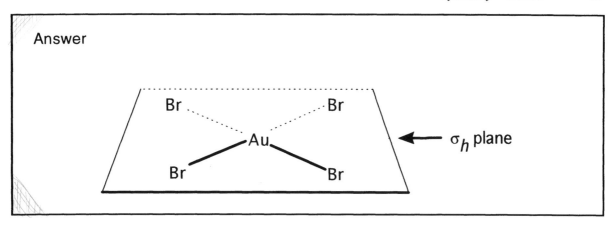

Answer

σ_h plane

The remaining reflection plane is shown above. This plane is *perpendicular* to the principal axis, and the molecule is completely contained within the plane. In fact, the symmetry operation defined by this reflection plane, on $AuBr_4^-$ would result in none of the atoms exchanging places. Despite this, it is still a symmetry element of the molecule. It is given the symbol σ_h, where '*h*' stands for 'horizontal'. It turns out that this is an important symmetry element to be able to identify. If you have had any problems 'seeing' this reflection plane, then it is worth building a simple molecular model of $AuBr_4^-$, so that you can see the molecule in three dimensions. The key thing to notice is that the σ_h plane is perpendicular to the principal axis.

2.4 Centres of inversion

We now come to another symmetry element which can be used to describe the symmetry of a molecule. This symmetry element is called a centre of inversion (or inversion centre) and it is given the symbol i. The good thing about this symmetry element is that, unlike rotation axes and reflection planes, a single molecule can only have one such centre, and we only need to decide whether a molecule has got one or not. The bad thing about it is that the actual symmetry element itself is sometimes difficult to 'see'. Certainly, this is one of the cases where practice makes perfect.

There are two ways to describe an inversion centre. Some people find it easier to understand one way rather than the other. The first way of 'seeing' an inversion centre is by inspection; this gets easier with practice. After a while, this is the quickest way of spotting an inversion centre. Take $AuBr_4^-$ as the example again. To 'see' the inversion centre in $AuBr_4^-$ by inspection, one has to imagine moving each atom in a straight line towards the Au atom, passing through the Au atom, and then moving away from the Au atom in the same direction to a point where the atom is at the same distance away from the Au atom as it was before the symmetry operation began. The inversion centre lies on the Au atom.

Sketch the $AuBr_4^-$ molecule and label each Br atom. Mark the position of the inversion centre with an arrow. Carry out the inversion operation as described above, and sketch the result.

Answer

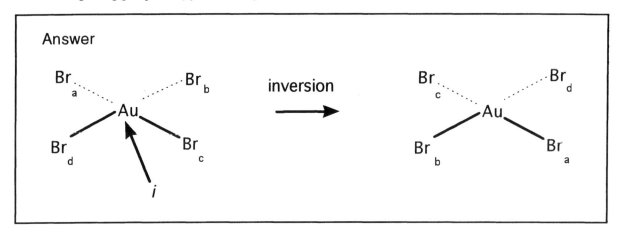

Of course, carrying out this operation on the Au atom does not change its position at all. The bromine atoms, however, are all interchanged with their 'opposite' bromine atoms. Since the final result after inversion is *indistinguishable* from the original, then $AuBr_4^-$ has a centre of inversion as a symmetry element.

The following two examples should also help in identifying an inversion centre, using the method described above.

For NH_3, (Z)-1,2-dichloroethene, (E)-1,2-dichloroethene and $CoCl_6^{3-}$ (octahedral), using the method described above, reason whether or not each molecule has an inversion centre of symmetry. If the molecule does have an inversion centre, label the atoms and sketch the result of the inversion operation.

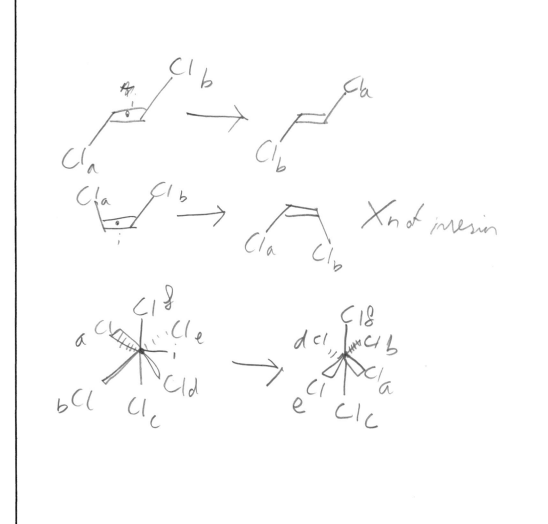

Answer

Only (*E*)-1,2-dichloroethene and $CoCl_6^{3-}$ have inversion centres of symmetry. The inversion centres are in the middle of the double bond and on the Co atom respectively. Carrying out the operations on these molecules is shown below.

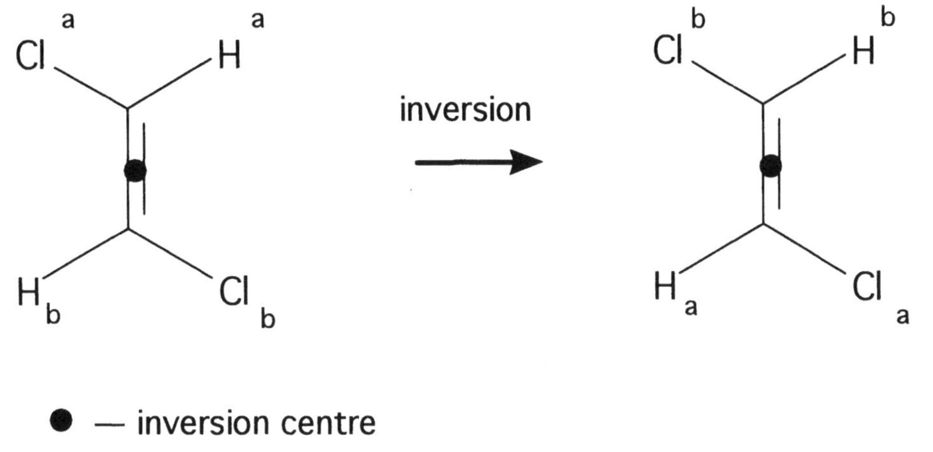

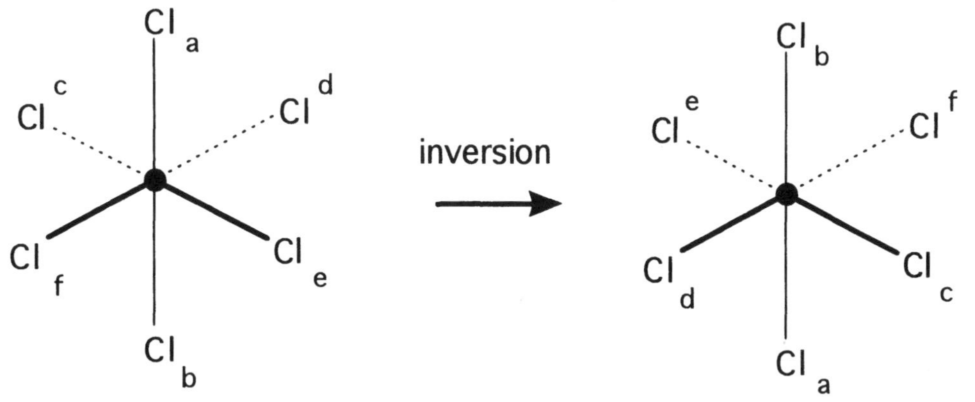

● — inversion centre

The second way to identify an inversion centre is to use a slightly more mathematical description. (If the mathematics worries you here, then you can use the qualitative definition given above, it works just as well. However, if possible, you should read this part anyway to see if you can understand it.) The mathematical description goes as follows: if the inversion centre is placed on the origin of a three-dimensional cartesian coordinate set of axes, and atoms are at general positions given by the coordinates (x, y, z), then an inversion centre is a symmetry element of that molecule if a molecule having the same atoms with coordinates $(-x, -y, -z)$ is indistinguishable from the original molecule. An example should help to make this clearer. The diagram at the top of page 16 shows the $AuBr_4^-$ molecule placed on a set of cartesian axes. The Au atom sits on the origin $(0, 0, 0)$. Each of the Br atom positions is given by a unique set of coordinates; so, for Br_a let us define the coordinates as (x_a, y_a, z_a). The inversion centre operation, with the symmetry element placed at the origin, transforms each of the coordinates as follows:-

$$(x, y, z) \xrightarrow{\ i\ } (-x, -y, -z)$$

For example, the Br_c atom coordinates after the original operation become $(-x_c, -y_c, -z_c)$. Since the new coordinates of each of the Br atoms have identical values to coordinates of one of the other Br atoms—so, for example $(-x_c, -y_c, -z_c) = (x_a, y_a, z_a)$—then the molecule after the symmetry operation is *indistinguishable* from the original. Therefore, the inversion centre is a symmetry element for $AuBr_4^-$.

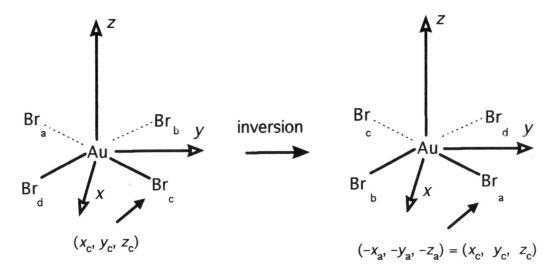

A third, but sometimes unreliable way is to realise that an inversion centre is exactly the same as the *combination* of a specific rotation axis and a specific reflection plane. This combination is rotation by 180° *followed by* reflection in a reflection plane perpendicular to the rotation axis. In other words, if a molecule contains a C_2 axis and a reflection plane perpendicular to this axis, then it must, by definition, also have an inversion centre of symmetry. (The C_2 axis need not be the principal axis.) The inversion centre element is placed at the meeting point of the axis and the reflection plane. But, BEWARE, the presence of an inversion centre of symmetry does not necessarily mean that the molecule has a C_2 axis and a perpendicular reflection plane as *individual* symmetry elements.

Using the methods on previous pages, reason whether or not the H_2O molecule has an inversion centre as a symmetry element? Do the same for $AuBr_4^-$. For $AuBr_4^-$, sketch the molecule and the key symmetry elements which help you decide whether it has an inversion centre or not.

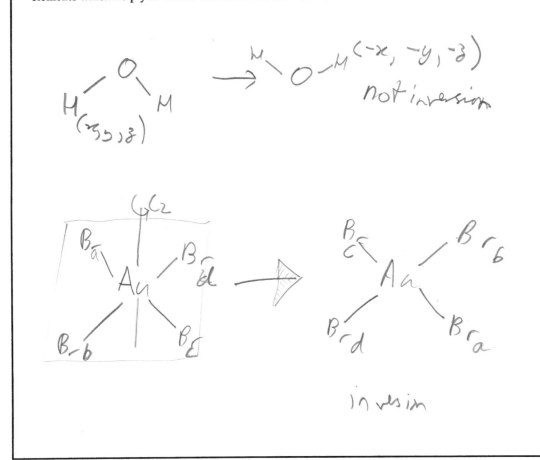

Answer

The H_2O molecule has a C_2 axis and two **mirror planes, but** neither of these mirror planes is perpendicular to the C_2 axis. Despite the fact that the molecule does not have a mirror plane perpendicular to a C_2 axis, we cannot say for definite that the molecule does not have an inversion centre of symmetry. We must use the other definitions to decide. Using the first definition: if, the origin is placed on the oxygen atom, and the hydrogen atoms are 'moved' towards the oxygen atom and then 'moved' away from the oxygen atom in the same direction to the same distance away from the O atom, then a different orientation of the water molecule is obtained, see figure. In fact, it is impossible to pick any origin where performing an inversion centre operation gives a molecule *indistinguishable* from the original. Therefore, the water molecule does not have an inversion centre of symmetry.

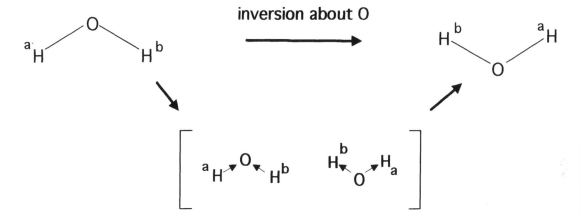

AuBr$_4^-$ does have a C_2 axis (coaxial with the principal C_4 axis) and it does have a reflection plane (same as σ_h) perpendicular to the C_2 axis. Therefore, the molecule also has an inversion centre.

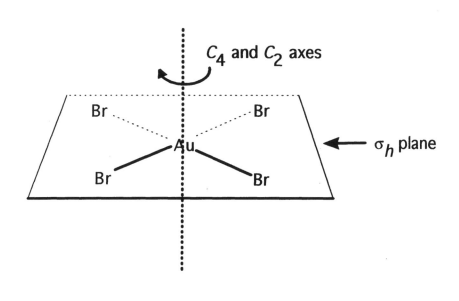

In the example above the inversion centre lies directly on the Au atom.

Which of the following molecules has an inversion centre of symmetry?

NO_3^-, SO_4^{2-}, H_2O_2, PCl$_5$, benzene, chair-form cyclohexane, boat-form cyclohexane, NiCl$_4^{2-}$, Ni(CN)$_4^{2-}$

Answer

H_2O_2 only has an inversion centre when the molecule is in a planar E (or *trans*) configuration. The inversion centre is then in the middle of the O–O bond.
Benzene—in the centre of the molecule.
Chair-form cyclohexane—in the centre of the molecule.
$Ni(CN)_4^{2-}$ (square planar)—on the Ni atom.

It is important that you can identify an inversion centre before you progress with the rest of this book, as it will be referred to again and again. If you have had difficulty in 'seeing' the inversion centre, then it is well worth building some of the molecules described in the question above, and seeing if you can find the inversion centre in three dimensions.

2.5   Improper rotation axes

This now brings us to the final symmetry element (and operation) that we need to know. I am often asked why this particular symmetry element is needed in group theory. The best way of answering this is to consider an example from a qualitative point of view. Consider the simple, tetrahedral molecule, CH_4. It is easy to see that all of the hydrogen atoms in this molecule are in identical chemical environments. In other words, the hydrogen atoms can be related by some *single* symmetry operation. That is, there must be a *single* symmetry operation that we can carry out which will transform each of the hydrogen atoms to all of the other hydrogen atom positions. This is exactly analogous to the C_2 axis in water relating the equivalent hydrogen atoms, and the C_3 axis in NH_3 relating *all* of the hydrogen atoms.

However, if one attempts to relate all of the hydrogen atoms in methane using our existing knowledge of rotation axes and reflection planes (the molecule does not contain an inversion centre of symmetry) it is impossible to find a *single* symmetry element which will allow us to transform each of the hydrogen atom positions into all of the other hydrogen atom positions. So, for example, the C_3 axis which is present as a symmetry element (see diagram below), describes a symmetry operation which transforms only three of the hydrogen atom positions onto each other; the fourth hydrogen atom on the rotation axis, cannot be transformed onto the other three positions by this operation. Clearly, we need another symmetry element which can describe a symmetry operation to relate all hydrogen atoms positions to each other.

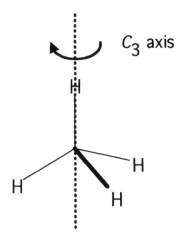

Enter the improper rotation axis. This is a single symmetry element, the operation of which can transform each of the hydrogen atom positions into all of the other hydrogen atom positions in CH_4. Unfortunately, this symmetry element is the most difficult to identify, and requires quite a bit of practice before one can be confident using it. As before, the best way of approaching this is to start with an example.

Consider methane again. The molecule has two types of rotation axes, C_3 and C_2. One of the C_3 axes is identified above.

Sketch the methane molecule, and identify the position of one of the C_2 axes of symmetry. (A small molecular model may be useful here.) Label the hydrogen atoms and sketch the effect of the C_2 symmetry operation.

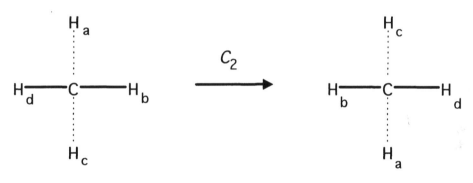

Answer

One of the C_2 axes can be seen if the molecule is viewed with the carbon atom in the plane of the paper, and two hydrogen atoms going into the plane of the paper, and two hydrogen atoms coming out of the plane of the paper. The view is looking right down the C_2 axis. Notice that the C_2 axis bisects the H–C–H angle.

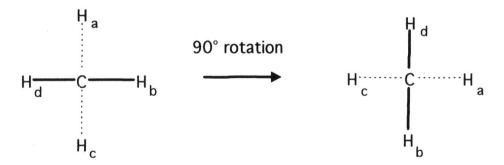

Using the C_2 axis, only H_a and H_c (or H_d and H_b) atoms can be interchanged. It is impossible to interchange the H_a and H_b atoms. For instance, a clockwise rotation of 90° (not a symmetry operation in itself for CH_4) does not give an indistinguishable result, as the H atoms are on opposite sides of the plane of the paper from where they would give an indistinguishable result (see figure below).

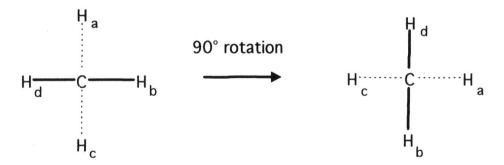

But, a rotation of 90° almost gives the right result. If we now combine the rotation by 90° with reflection in a reflection plane perpendicular to the axis of rotation (so, in the figure the reflection plane is in the plane of the paper), then we obtain a molecule which is *indistinguishable* from the original (see figure below). In fact, this *combination* of operations gives us a single operation which is a *symmetry operation* for the molecule. Note, however, that the molecule does not necessarily have a either C_4 axis of rotation, or a reflection plane as individual symmetry elements. (This is analogous to the combination of C_2 followed by reflection in a reflection plane for an inversion centre of symmetry. The presence of an inversion centre does not necessarily imply that there are individual C_2 and reflection plane elements of symmetry.)

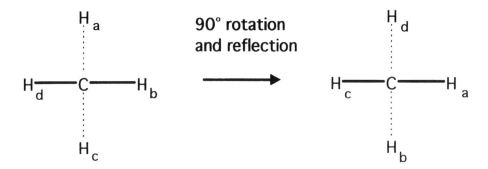

The combination of a rotation and reflection in a perpendicular reflection plane gives a new symmetry element, which is called an improper rotation axis. In the example given above, this axis is given the symbol S_4, because the rotation is 90°, and 360/90 = 4.

   Before we move on to try some examples, there is a couple of important points about improper rotation axes. First, we need to study the effects of successive applications of an improper rotation on a molecule. For instance, take the methane example above; what is the effect of applying the S_4 operation once, twice, three times or four times in succession? As with the rotation axes, we can use the following notation to represent successive operations: one operation = S_4^1, two operations S_4^2, three operations S_4^3 and so on. The S_4^1 operation is shown in the figure above.

Using the methane example shown above, sketch the molecule with labelled hydrogen atoms and show the effect of S_4^2 and S_4^3 operations.

Answer

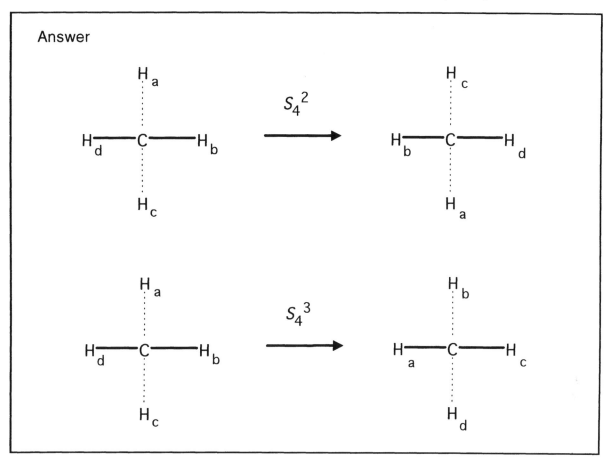

The S_4^2 operation is exactly the same as the C_2^1 operation, and the S_4^4 operation is the same as the E operation. Therefore, the only *unique* symmetry operations from an S_4 axis are S_4^1 and S_4^3. Notice, therefore, that an S_4 axis must have a coaxial C_2 axis as a symmetry element in the molecule. In fact, without taking the point much further, it is true that for any even-order S_n axis, there must be a coaxial $C_{n/2}$ axis.

For an S_6 axis, the successive application of the improper rotation also leads to operations which can be described by other symmetry operations.

Perform successive S_6 operations on a general molecule and figure out which operations can be described by other symmetry operations, e.g. S_6^2 is the same as C_3^1

Answer

S_6^2 is the same as C_3^1
S_6^3 is the same as i
S_6^4 is the same as C_3^2
S_6^6 is the same as E

Only S_6^1 and S_6^5 are unique

The S_2 operation is unusual insofar as it is exactly the same as carrying out an inversion operation, since it corresponds to a rotation by 180° followed by reflection in a perpendicular reflection plane. (See the section on centres of inversion.) Accordingly, the S_2 operation does not concern us as chemists.

Improper rotation axes with odd orders (e.g. $S_3, S_5, S_7...$) require a little more care. Taking S_3 as an example. The problem here is that the S_3^3 operation is *not* the same as the E operation. In fact, the S_3^3 operation corresponds to a σ operation. You can 'see' this if you realise that performing the improper rotation an odd number of times, means that an odd number of reflections has been performed, which necessarily means that the result of S_3^3 is a reflection in a plane perpendicular to the improper rotation axis. However, the S_3^6

operation (i.e. a double rotation) is the same as the E operation, since now an even number of reflections has been applied. Without going into the details of which other symmetry operations are the same as S_3^n operations, the corresponding operations are listed below.

$$S_3^2 \text{ is the same as } C_3^2$$
$$S_3^3 \text{ is the same as } \sigma$$
$$S_3^4 \text{ is the same as } C_3$$
$$S_3^6 \text{ is the same as } E$$

Only S_3^1 and S_3^5 are unique symmetry operations.

A similar argument applies to other odd-order improper rotation axes. These are not described here, but the 'take-home message' is that many of the improper rotation operations can be described by other symmetry operations, and that with odd-order improper rotation axes, S_n^{2n} is the same as E.

Now try the following examples:-

PCl_5 has a trigonal bipyramidal structure. Sketch the molecule and identify the improper rotation axis. By labelling the fluorine atoms show the effect of the S operation. (Hint, first identify any rotation axes.)

Answer

The S_3 axis is coaxial with the C_3 axis.

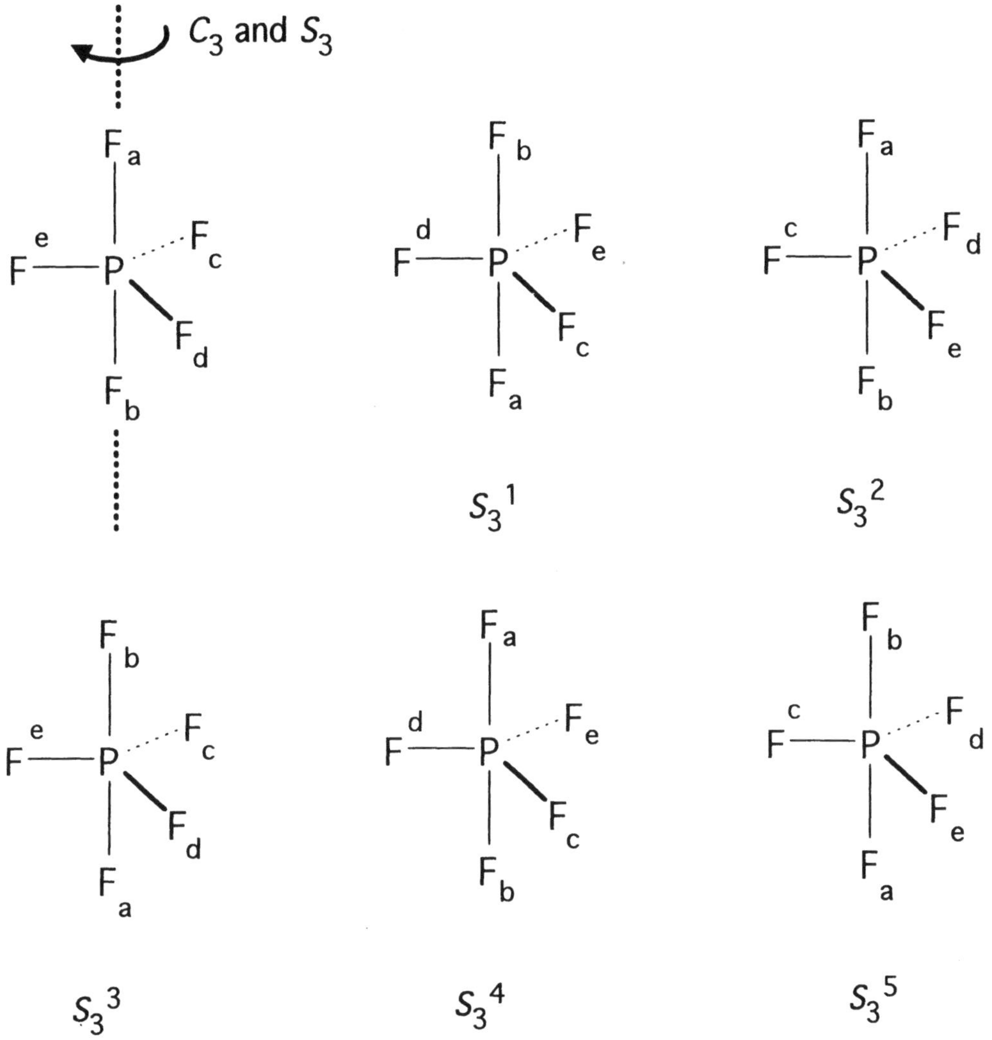

C_3 and S_3

S_3^1

S_3^2

S_3^3

S_3^4

S_3^5

Identify the S_6 axis in $CoCl_6^{3-}$ (an octahedral complex). This can be a difficult axis to 'see'. This is an even-order improper rotation axis and it must have a coaxial $C_{n/2}$ axis. Therefore, it is worth looking for the corresponding C_3 axis first. A molecular model is useful for this example. Sketch the molecule viewed down the C_3 axis, and by labelling the chlorine atoms, show the effect of an S_6^1 operation.

Answer

If the model is viewed in the correct orientation, then the S_6 axis can be seen quite clearly. See the diagram below, which is viewed directly down the S_6 (and C_3) axis. The effect of a $S_6{}^1$ operation is shown.

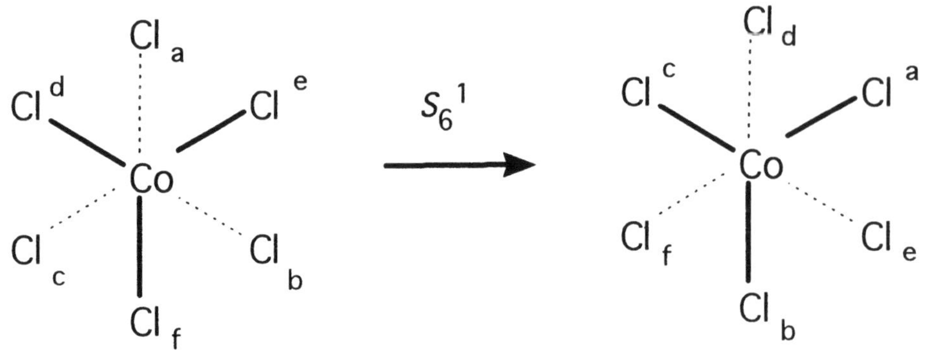

This brings us to the end of the symmetry elements and operations which are used to describe the symmetry of molecules. You can now describe the symmetry of any molecule that you are presented with in terms of its symmetry elements. This process of identifying symmetry elements gets easier with practice, and it is worth persevering with, until you are confident in 'seeing' all of the symmetry elements and operations. The examples below are designed to give you practice at identifying symmetry elements. Before attempting these examples make sure that you understand what each of the symmetry operations does. The symmetry elements that we have studied are:

- Rotation axes (symbol C_n)
- Reflection planes (symbol σ)
- Inversion centres (symbol i)
- Improper rotations (symbol S_n)

Identify and list as many symmetry elements as you can in the following molecules:

PCl_5 (trigonal bipyramidal)
(*E*)-1,2-dibromoethane
CH_4

Answer

PCl_5 has the following symmetry elements: E, C_3, $3C_2$, σ_h, S_3 and $3\sigma_v$. (More completely, there are two C_3 operations, which are $C_3{}^1$ and $C_3{}^2$, and there are two S_3 operations $S_3{}^1$ and $S_3{}^5$.)

(*E*)-1,2-dibromoethane has the following symmetry elements: E, C_2, i and σ_h.

CH_4 has the following symmetry elements: E, C_3, C_2, S_4 and σ_d (these are dihedral planes since they bisect the H–C–H angles). There are several of each of the symmetry elements except E.

2.6   Successive operations, identity, inverse and class

Before we go further it is important that we look at some of the properties of symmetry operations in a molecule. These properties will be required in section 3, where we will build on them to develop further our understanding of group theory.

    In this part we will study some of the properties of symmetry operations. Most of the results we come across here simply have to be remembered, they do not have to be proved.

2.6.1   Successive operations

The first question to ask is: what happens if we carry out successive symmetry operations? We have already seen the successive combination of rotations with reflections to describe improper rotations (rotation followed by reflection), but what about any combination of symmetry operations; for example, is there any difference between a rotation followed by a reflection and a reflection followed by a rotation? In other words, is the order in which we perform operations important? Again, this is probably best illustrated by an example.

Phosphine, PH_3, has a pyramidal structure like NH_3. Sketch the structure and identify the symmetry elements of the structure.

Answer

PH_3 has E, C_3 and $3\sigma_v$ symmetry elements.

Again, for PH_3, sketch the structure, and by labelling the hydrogen atoms show the effect of performing the following combinations of symmetry operations. (Use the same orientation of reflection plane in each case.)

a) σ_v after $C_3^{\,1}$.
b) $C_3^{\,1}$ after σ_v.

Answer

The PH_3 molecule is sketched below with labels on the hydrogen atoms. The reflection plane is the one that lies in the plane of the paper. The top scheme shows σ_v performed after C_3^1 and the bottom scheme shows C_3^1 performed after σ_v.

The exercise shows that the result of performing symmetry operations in a different order does *not* give identical results. Indeed, this is a general rule for combining symmetry operations: the order of combination *is* important. Taking this point a little further, with the above example we have shown that:

C_3^1 performed after σ_v does not give the same result as σ_v performed after C_3^1.

This can be expressed in a more mathematical way. If we replace the 'performed after' with a multiplication sign, then we get.

C_3^1 performed after σ_v can be written as: $C_3^1 \times \sigma_v$

and

σ_v performed after C_3^1 can be written as: $\sigma_v \times C_3^1$

If we say that 'same as' is replaced by an equals sign '=', and 'not the same as' is replaced by a not equals sign '$\neq$' —also, let us dispense with the multiplication sign (as in normal algebra)—then we can write the following for this example (for other examples this may not necessarily be true):

$$C_3^1\, \sigma_v \;\neq\; \sigma_v\, C_3^1$$

There are two things to remember about this particular result. First, we have expressed the symmetry operations in a more mathematical way. (Notice here that, in the equation, the operation which is performed second is written first. This may sound like an odd way of doing things, but it is easier to understand if the symmetry operation symbol is read as 'operates on whatever is to the right'. So, for example, $C_3^1\, \sigma_v$ can be read as C_3^1 operates on the result of σ_v. This way of writing symmetry operations takes some getting used to.)

Second, for the mathematicians, this general property of the order of performing an operation being important is called a *non-commutative* property. We will come back to this in the next section, and it simply needs to be remembered for the moment.

So, we have seen that the order of performing symmetry operations is important. The combination of symmetry elements also leads to the following important result. The result of any combination of symmetry operations can be described by a single symmetry operation of the molecule. So, for the example above we can say the following:

$$\sigma_v \, C_3{}^1 = \; \sigma_v{}'$$

and

$$C_3{}^1 \, \sigma_v \; = \; \sigma_v{}''$$

Where $\sigma_v{}'$ and $\sigma_v{}''$ are the results of the symmetry operations of the other reflection planes in the molecule. The effect of $\sigma_v \, C_3{}^1$ is illustrated below (remember that the first operation to perform is the one on the right).

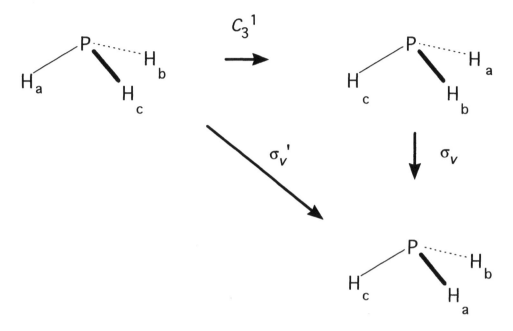

The diagram shows that the operation of $C_3{}^1$ followed by reflection in σ_v gives an identical result to reflection in $\sigma_v{}'$ ($\sigma_v{}'$ is the reflection plane which contains the *original* P–H$_b$ bond and bisects the H$_a$–P–H$_c$ angle). In fact, it is a general result of symmetry operations of a molecule that any combination of symmetry operations within a single molecule is the same as performing a single symmetry operation. This will be left as a general result to be remembered—we will need this result in the next section. If you are in any doubt about the truth of the statement, try a few examples for yourself; PCl$_5$ is a good molecule to try in this respect.

So far we have combined two symmetry operations, what about performing three successive symmetry operations?

For the PH$_3$ example given above, what is the effect of combining the following operations? Remember that σ_v is the reflection plane which lies in the plane of the paper, and $\sigma_v{}'$ is the reflection plane which contains the *original* P–H$_b$ bond before any symmetry operations are carried out (i.e. this reflection plane does not 'move' with the P–H$_b$ bond during the operations). The parentheses in the last two examples mean that the combination of the operations in the parentheses must be carried out before any other combinations, following the same rule when combining two operations that the one on the right is performed first. (Hint: with the parentheses, first figure out what single symmetry operation is the same as the combination of the two operations within the parentheses). It will probably help to sketch the molecule with labelled hydrogen atoms in each instance.

$C_3{}^1 \, \sigma_v{}' \, \sigma_v$

$\sigma_v{}' \, \sigma_v \, C_3{}^1$

$(C_3{}^1 \, \sigma_v) \, \sigma_v{}'$

$C_3{}^1 \, (\sigma_v \, \sigma_v{}')$

Answer

$C_3{}^1 \sigma_v{}' \sigma_v$

Splitting this up. We first need to figure out what single operation $\sigma_v{}' \sigma_v$ corresponds to. This can be done by sketching out the molecule as follows:

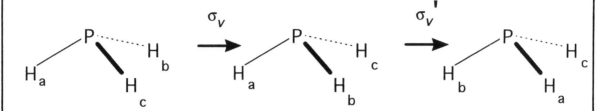

From the figure, it can seen that $\sigma_v{}' \sigma_v = C_3{}^2$.
Therefore, we can say that $C_3{}^1 \sigma_v{}' \sigma_v = C_3{}^1 C_3{}^2$, which can similarly be worked out as $= E$. So, the answer is:-

$C_3{}^1 \sigma_v{}' \sigma_v = E$

The other answers are:-

$\sigma_v{}' \sigma_v C_3{}^1 \;=\; C_3{}^2 C_3{}^1 = E$

$(C_3{}^1 \sigma_v) \sigma_v{}' \;=\; \sigma_v{}'' \sigma_v{}' \;=\; C_3{}^2$

$C_3{}^1 (\sigma_v \sigma_v{}') \;=\; C_3{}^1 C_3{}^1 \;=\; C_3{}^2$

Although the answers above are for a specific case, they illustrate two important, and, as it turns out, general properties of symmetry operations in molecules. First, the order in which operations are combined is still important—i.e. they are non-commutative. Second, the last two examples show that the same answer is obtained ($C_3{}^2$ in this case) despite working out different parts in a different order. In general, this means that the following is true (where A, B and C denote any particular symmetry operation):

$$A(BC) \;=\; (AB)C$$

To mathematicians, this means that the combination of symmetry operations is *associative*. This is simply a name for this type of mathematical behaviour, and just needs to be remembered for the moment.

2.6.2 Identity

We have already established that the operation of doing nothing, E, is an important part of describing the symmetry of a molecule. For instance, E can be seen as the result of combinations of other symmetry operations. This fact, again, needs to be remembered for now, as we will build on it in the next section.

2.6.3 Inverse

Another property of symmetry elements is that the following is always true (where A is any particular symmetry operation and A^{-1} is also a symmetry operation of the molecule):

$$A^{-1}A \;=\; E$$

What does this mean in words? It says that any symmetry operation (A) performed on a molecule, can be reversed, or undone, by another symmetry operation of the molecule, to give back the original molecule. The A^{-1} operation is called the *inverse* of the A operation. For example, in the PH_3 example above, each of the operations has an inverse.

Write the inverse of the following symmetry operations in PH_3.

E, $C_3{}^1$, $C_3{}^2$, σ_v.

Answer

E has inverse E (it is its own inverse).
$C_3{}^1$ has inverse $C_3{}^2$
$C_3{}^2$ has inverse $C_3{}^1$
σ_v has inverse σ_v (again, it is its own inverse).

This result, that each symmetry operation has an inverse operation which is also a symmetry operation of the molecule is a general result and we will use this in the next section.

2.6.4 Class

By now, you are hopefully familiar with the notion that the symmetry of a molecule can be expressed in terms of a collection of symmetry operations which can be performed on the molecule. These symmetry operations can then be used to help define the symmetry of the molecule. For example, the symmetry operations in a molecule like NH_3 are E, $C_3{}^1$, $C_3{}^2$, σ_v, $\sigma_v{}'$ and $\sigma_v{}''$. From a qualitative point of view (not a proof), several of these symmetry operations are similar in the symmetry operations that they perform. $C_3{}^1$ and $C_3{}^2$ are similar, and σ_v, $\sigma_v{}'$ and $\sigma_v{}''$ are also similar. E is by itself.

It turns out that these 'similar' operations can be related by a mathematical procedure, called a *similarity transformation*. You do not need to know anything about this transformation, but what you do need to know is that similar symmetry operations can be grouped into *classes*. So, NH_3 (and indeed, any molecule which has the same symmetry operations) can be said to have the following *classes* of symmetry operations:- E, $2C_3$ and $3\sigma_v$. Where the '2' and the '3' denote how many individual operations are in that class—$C_3{}^1$ and $C_3{}^2$ are in the $2C_3$ class. (Note that this qualitative 'rule-of-thumb' is not always true.)

Putting symmetry operations into classes is a means of avoiding having to write out all of the symmetry operations. From a mathematical point of view, classes are useful since symmetry operations within a class also have similar mathematical properties. For now this needs to be remembered, and we will make use of it in the next section.

2.7   Point groups

This is now the last part of this section. Here's a quick recap on what we have covered so far:

- Rotation axes, reflection planes, inversion centres and improper rotation axes are all symmetry elements which describe particular symmetry operations.
- The symmetry of every molecule can be described in terms of its symmetry operations.
- The symmetry operations can be combined following certain rules, to give a mathematical-like behaviour.

These facts now provide a foundation for looking at the symmetry of molecules in more detail, and we will make repeated use of them. The only problem of using these repeatedly is that it can be cumbersome to have to write out all of the symmetry operations of a molecule every time we want to study its symmetry. Fortunately, there is a shorthand way of representing the symmetry of molecules. In this last part we will go over how this shorthand method works.

List all of the unique symmetry operations that you can find in a PCl$_5$ molecule (trigonal bipyramidal). It will help to sketch the molecule, and a small molecular model may also help.

Answer

The symmetry operations are:

$$E, C_3{}^1, C_3{}^2, C_2\ C_2{}'\ C_2{}'', \sigma_h, S_3{}^1\ S_3{}^5, \sigma_v\ \sigma_v{}'\ \sigma_v{}''$$

Notice that the redundant symmetry operations, such as $S_3{}^2$ which is the same as $C_3{}^1$, have been missed out. The symmetry elements are shown below.

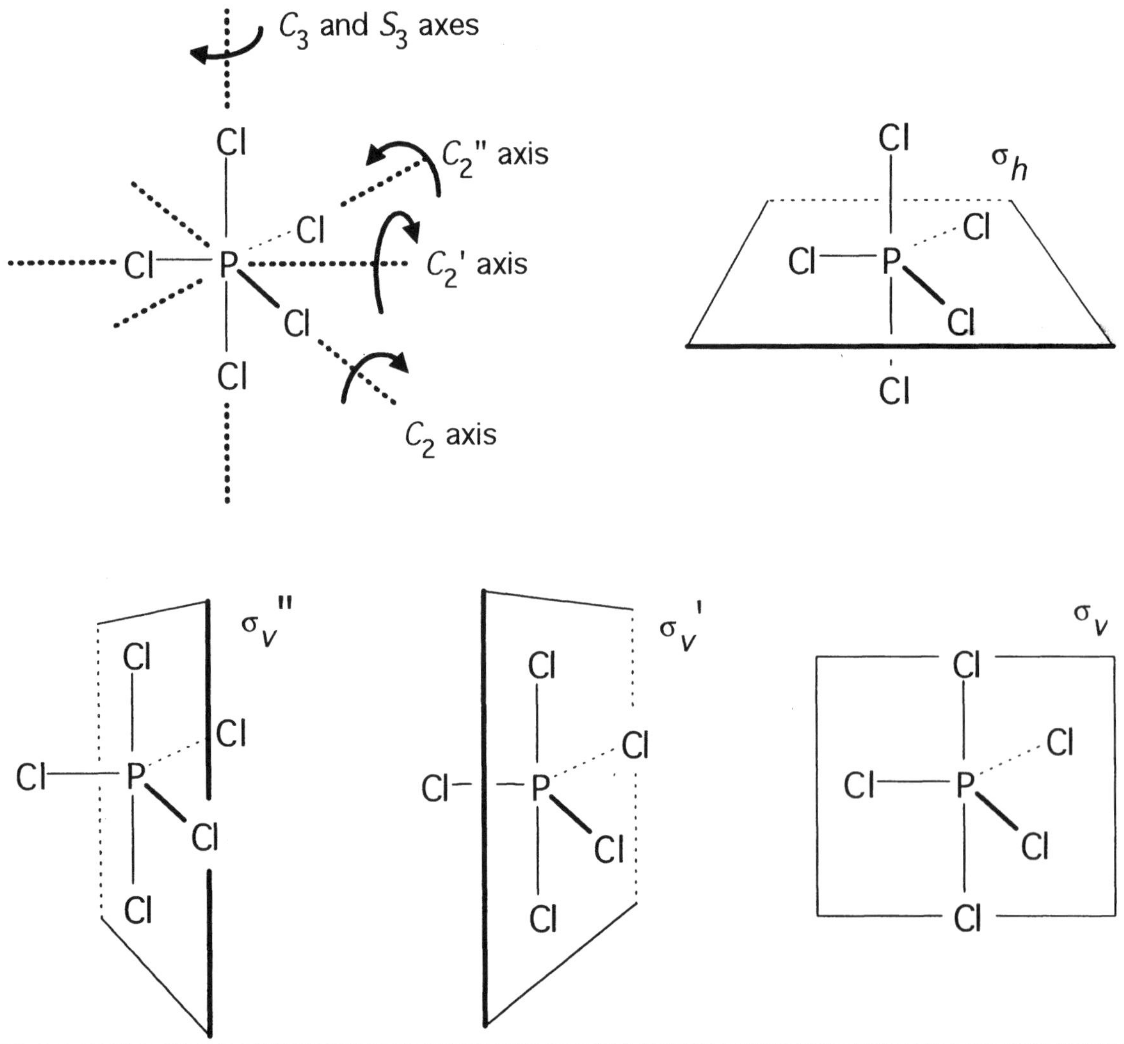

To describe the symmetry of a molecule by listing all of its symmetry operations is clearly too time-consuming. Therefore, the following classification system is used. It relies on being able to identify key symmetry elements within a molecule. The key symmetry elements then define a particular group, which has several different symmetry elements. So, for example, to classify the PCl_5 molecule, it is only necessary to identify the principal axis of rotation, the perpendicular C_2 axes of rotation and the σ_h reflection plane. This takes away the drudgery of having to identify every single symmetry operation. All of these classifications have been worked out beforehand, and all we need to do is to follow the rules for getting to the particular classification (see below).

Each classification is given a symbol, which uniquely identifies that classification. Each symbol stands for a collection of symmetry operations. The symbol also represents what is called a *point group*. The two words *point* and *group* each mean something. *Point* means that all of the symmetry elements associated with the symmetry operations pass through a single point in space. This point is not changed in position by any of the symmetry operations. For example, the 'point' in the PCl_5 molecule lies directly on the phosphorus atom. (Beware, however, it is not necessary for the 'point' of the group to lie on an atom, e.g. H_2O_2.) The word *group* means that we have a group of symmetry operations. We shall see in the next section that we can define a group mathematically, but for the moment it is sufficient to recognise that we are dealing with a group of symmetry operations in each *point group*.

We classify a molecule into a point group by answering some simple questions about the molecule.

Question 1.
Is the molecule is one of the following 'recognisable' groups

NO: Go to question 2
YES: octahedral, given the point group symbol O_h
tetrahedral, given the point group symbol T_d
linear, given the point group symbol $C_{\infty v}$ (if it does not have an i symmetry element)
linear, given the point group symbol $D_{\infty h}$ (if it also has an i symmetry element)

Question 2.
Does the molecule possess a rotation axis of order ≥ 2?

YES: Go to question 3
NO: If it has no other symmetry elements, then it is given the point group symbol C_1
If it has one reflection plane of symmetry, then it is given the point group symbol C_s
If it has a centre of inversion, then it is given the point group symbol C_i

Question 3
Has the molecule more than one rotation axis?

YES: Go to part 4.
NO: If it has no other symmetry elements, then it is given the point group symbol C_n
(where n = the order of the principal axis, e.g. C_3).
If it also has one σ_h, then it is given the point group symbol C_{nh}
(where n = the order of the principal axis, e.g. C_{2h}).
If it has n σ_v, then it is given the point group symbol C_{nv}
(where n = the order of the principal axis, e.g. C_{3v}).
If it has an S_{2n} axis coaxial with the principal axis, then it is given the point group symbol S_{2n}

Part 4
The molecule can be assigned a point group as follows:

If it has no other symmetry elements, then it is given the point group symbol D_n
(where n = the order of the principal axis, e.g. D_3).
If it has got n σ_d reflection planes bisecting the C_2 axes, then it is given the point group symbol D_{nd}
(where n = the order of the principal axis, e.g. D_{4d}).
If it also has one σ_h, then it is given the point group symbol D_{nh}
(where n = the order of the principal axis, e.g. D_{3h}).

This series of questions allows us to identify all of the common point groups encountered in chemistry. There are other possible point groups, but fortunately these are rare. As a starting point is it important that you can recognise some specific point groups; these are the O_h (for octahedral), T_d (for tetrahedral), $C_{\infty v}$ for a linear molecule without a centre of symmetry (e.g. HCN) and $D_{\infty h}$ for a linear molecule with a centre of symmetry (e.g. CO_2). Hopefully all of these cases are easily recognisable, and it should be a fairly simple exercise in assigning the point groups. Note, however, that for a molecule to be in the O_h or the T_d point group it must be *perfectly* octahedral or tetrahedral respectively, if it is not then it should be assigned to a different point group. Identifying these groups quickly gets better with practice, and the exercises below should help.

Using the questions above, assign the point group of PCl_5. It may help to refer back to the symmetry elements in PCl_5 that are shown in the previous exercise.

Using the questions above, assign the point group of NH_3.

Answer

PCl$_5$

Question 1.
PCl_5 has a trigonal bipyramidal shape and does not fall into any of the special point groups.

Question 2.
YES: there is a C_3 axis which passes through the linear F–P–F part of the molecule.

Question 3.
YES: there are C_2 axes which are perpendicular to the principal axis.

Question 4.
The molecule has several other symmetry elements. It has three σ_v planes. However, it also has a σ_h plane. Therefore, it is assigned to point group D_{3h}.

NH$_3$

Question 1.
NH_3 has a pyramidal shape and does not fall into any of the special point groups.

Question 2.
YES: there is a C_3 axis.

Question 3.
NO: there are no other rotation axes other than the principal C_3 axis.
It does have other symmetry elements. It does not have a σ_h plane, but it does have three σ_v planes. Therefore, it is assigned to point group C_{3v}.

The point groups are shorthand for the symmetry elements within a molecule. For example, if a molecule is assigned to a C_{3v} point group, then it has *all* of the following symmetry elements: $E, 2C_3, 3\sigma_v$. It is a very important part of group theory in chemistry that you can assign the point group of a molecule accurately. It is definitely worth spending time practising this aspect of the subject until you are confident of assigning correct point group each time. Try to assign the point groups for all of the molecules in the exercise below.

Assign point groups to the following molecules:

H_2O, PH_3, SO_2, HCl, $AuBr_4^-$, $CoCl_6^{3-}$, (*E*)–1,2–dibromoethane), benzene, methylbenzene, trichloromethane, NO_3^-, SO_4^{2-}, $HCCH$, B_2H_6, $Co(en)_3^{3+}$ (en = 1,2–diaminoethane), $^-O_2CCH_2NH_3^+$ (glycine).

Answer

H_2O (**C_{2v}**), PH_3 (**C_{3v}**), SO_2 (**C_{2v}**), HCl (**$D_{\infty h}$**), $AuBr_4^-$ (**D_{4h}**), $CoCl_6^{3-}$ (**O_h**), (*E*)–1,2–dibromoethane (**C_{2h}**), benzene (**D_{6h}**), methylbenzene (**C_{2v}**), trichloromethane (**C_{3v}**), NO_3^- (**D_{3h}**), SO_4^{2-} (**T_d**), $HCCH$ (**$D_{\infty h}$**), B_2H_6 (**D_{2h}**), $Co(en)_3^{3+}$ (**D_3**), glycine (depends on the conformation, but if the ^-O_2CCN part is all planar, and one of the hydrogen atoms of the ammonium part is also in this plane, then the molecule has a single reflection plane and it is assigned **C_s**, otherwise it is **C_1**.)

2.8   Summary

In this section we have seen that our qualitative understanding of the symmetry of molecules can be made more quantitative by determining how the molecule is symmetric with respect to several symmetry operations. Each molecule can be said to contain one or more symmetry elements, which completely describe the symmetry of the molecule. Molecules can also be classified in point groups. A point group is simply shorthand for the collection of symmetry elements that describe the symmetry of a molecule. We can determine the point group of a molecule by answering several questions about the symmetry elements of the molecule.

We are now in a position where we can classify molecules according to their symmetry and also describe the symmetry elements of a molecule. The symmetry operations associated with these symmetry elements appear to follow mathematical rules like multiplication. The question now is can we make the transition into fully representing the symmetry operations of a molecule mathematically? If we can, can we then make quantitative predictions about how symmetry affects various energy levels in the molecule? The next section shows how this can be done.

What are the main points of this section?

- The energy levels of a molecule are linked, in some way, to the symmetry of the molecule.

- The symmetry of a molecule can be described with symmetry elements.

- Symmetry operations, described by symmetry elements, can be carried out on a molecule.

- Rotation axes, reflection planes, inversion centres and improper rotation axes are symmetry elements.

- The symmetry operations of a molecule have mathematical-like properties.

- Each set of symmetry operations has an inverse and identity.

- The combination of symmetry operations on a molecule is non-commutative, but it is associative.

- Every molecule can be classified into a point group which completely describes it symmetry.

SECTION 3

What is a group?

What is a group?

3.1   What is a group?

If you were to ask yourself 'what is a group' before any group theory course, then you might say something like 'a collection of persons and/or objects bound together by something in common' This is most people's understanding of a group. We might say that the symmetry operations of a molecule form a group in this sense. But clearly, if we are to make any headway at all in understanding how to quantify a group we must first define what we mean by a group much more rigorously, and, if possible, use mathematics to describe our group.

This short section aims to put across the idea of a mathematical group. But, before all of the non-mathematicians immediately close the book, the amount of mathematics needed here is nothing more than very simple addition and multiplication, and everyone should be able to come to grips with the arithmetic. It is, nonetheless, a very important section to read and understand, because much of the reason why group theory works in chemistry and much of the terminology will be more easily understood after reading this section. The more serious mathematics comes in the next section.

3.2   An example of a group

Many books on chemical group theory start their explanation of a mathematical group with molecules (usually the water molecule). However, it is useful to realise that the notion of a mathematical group extends to many examples other than symmetry in molecules. Accordingly, we will start with a non-chemical example, which lends itself nicely to explaining the origins of a mathematical group.

Consider a single soldier on parade. This soldier has to obey any one of four commands. These are: do nothing (!), right turn, left turn and about turn. Let us give each of these *operations* a symbol. Do nothing —E, turn right—R, turn left—L and about turn—A. We can represent the orientation of the soldier with an arrow, as in the figure below.

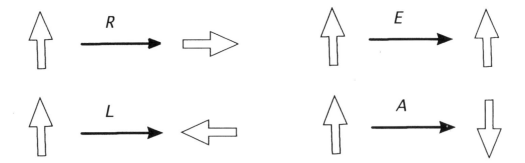

What is the result of giving several orders in succession? Right turn followed by right turn is the same as about turn. Left turn followed by right turn is the same as doing nothing. As in the previous section, we can represent these combinations of operations in a more mathematical sense.

Using the method in the previous section, represent the following in a more mathematical way.

Left turn followed by right turn is the same as doing nothing.
About turn after right turn is the same as left turn.
About turn before right turn is the same as left turn.
A combination of right turn followed by about turn before a right turn is the same as doing nothing.
About turn after the combination of a right turn with another right turn is the same as doing nothing.
About turn followed by left turn all after a right turn followed by a right turn is the same as left turn.

Answer

$RL = E$
$AR = L$
$RA = L$
$R(AR) = RL = E$
$A(RR) = AA = E$
$(LA)(RR) = RA = L$

Remember that the operation done last is on the left.

You should already be able to see similarities between the *operations* of right turn etc. with the symmetry *operations* that we discussed in the previous section. For example, successive operations can always be described by a single operation and the operations can be expressed more mathematically as shown with expressions in the exercise. Can we draw more similarities between the two types of operation? The answer is yes.

First, there is an identity operation that itself can be the result of combining other operations. Second, the combination of operations is associative as shown by a couple of the examples above. (This turns out to be general for all similar groups.) Third, the example above is actually commutative, however this is not a general case; usually such groups are not commutative.* Fourth, there is an inverse operation for all of the commands, e.g. R is the inverse of L, A is its own inverse.

3.3 Properties of a group

Therefore, there are very definite parallels between the symmetry operations of a molecule and the operations carried out by the soldier on parade. Mathematicians recognised the existence of such 'groups' a long time ago, and wrote down a set of rules which define what a mathematical group is. Unfortunately, the mathematicians decided to call the group members 'elements'. The elements of a group must not be confused with the symmetry elements, both are quite different. The elements of the groups that are important in chemistry are the symmetry *operations* of the molecule. The definition of a 'mathematical group' is as follows:

1. The combination of any two of the group elements must be the same as a single group element (we have already seen this with the symmetry operations of a molecule).

2. One element in the group must do nothing—the identity element.

---

* It is possible to see that the soldier example is non-commutative if you throw in the extra commands 'stand on your head' symbol I, and 'turn right with standing on your head' symbol I_R, and 'turn left with standing on your head' symbol I_L, and 'do a back flip to face the other way standing on your head', symbol I_A. All of these commands together with R, L, A and E form a group. It can be shown that $IR = I_R$, and $RI = I_L$, which, of course is non-commutative. The reason why they are non-commutative is that if you issue a command to the soldier who is upside down to turn right, he/she will appear to turn left. Try this at home if you are still unsure! With the extra 'invert' operations, the operations are still associative, but no longer commutative.

3. The combinations of elements must be associative. In other words, this general case must be true:

$$A(BC) = (AB)C$$

4. Every element must have an inverse (i.e. something to undo what the element does) which is also an element of the group. For example, for the soldier on parade, the inverse operation of R is L and vice versa, with both elements as part of the group.

It is probably already apparent that the symmetry operations of a molecule, covered in the previous section, follow the rules of a mathematical group. If in any doubt, turn back to section 2.7. "What good is this?" you may ask. The answer is that mathematical groups have unusual mathematical properties. In particular, the mathematics of group theory allow us to simplify greatly the equations used to determine possible wavefunctions (and energies) of a molecule. Therefore, by applying the mathematics of group theory in the context of molecular symmetry, it becomes possible to work out easily the possible energy levels of a molecule. More of this later, when we apply group theory to chemical problems, but it is important to realise that it is the powerful mathematics associated with a mathematical group which makes group theory such an important tool in chemistry.

Our job now is to marry the mathematics with the chemistry. Whilst this may sound like an almost impossible task, most of the difficult concepts have already been covered, and only the most basic knowledge of arithmetic is needed to go further.

3.4   Representations of a group

Let us look back at our soldier on parade. How can we represent the properties of this group more mathematically? We can start by looking at how we combine operations. We already know that we can combine operations to give an overall operation. For instance:

$$RR = A$$

A full picture of all combinations can be obtained if a *multiplication table* is prepared. One of these is given below. There is a row and a column to the table, with the group elements (i.e. the operations of the soldier) listed in order.

	E	R	L	A
E				
R				
L				
A				

The entries to the table are given by the combination of elements, with the element on the top row being the element on the right in the combination. So, we can fill out the first row of the multiplication table as follows:

	E	R	L	A
E	EE	ER	EL	EA
R				
L				
A				

Fill out the remaining entries yourself.

Answer

	E	R	L	A
E	EE	ER	EL	EA
R	RE	RR	RL	RA
L	LE	LR	LL	LA
A	AE	AR	AL	AA

Now draw up another table like above, but replacing the two letter combinations with the single letter equivalent. For example replace *EE* with *E*.

Answer

	E	R	L	A
E	E	R	L	A
R	R	A	E	L
L	L	E	A	R
A	A	L	R	E

The multiplication table now contains the single operation results of the combination of operations.

We now arrive at an important question. Is it possible to *represent* the effects of combining the operations with something more 'mathematical'? In other words, can we use normal mathematics to mimic the multiplication table given above, and still conform to the properties of a mathematical group?

The starting point is to replace the act of combining the operations with a multiplication sign, as we did for the symmetry operations in the previous section. But if we are to make headway we must eventually multiply numbers together to make it a truly mathematical representation. The simplest number we can replace each operation with is 1. What happens when we replace each of the operations with the number 1? Well we can draw up the original and 'mathematical mimic' multiplication tables:

	E	R	L	A
E	E	R	L	A
R	R	A	E	L
L	L	E	A	R
A	A	L	R	E

	1	1	1	1
1	1	1	1	1
1	1	1	1	1
1	1	1	1	1
1	1	1	1	1

The table on the left is the original table, with the results of operation combination. The table on the right is the mathematical mimic of the original table, where we have replaced each of the operations with a number 1, and said that the combination of the numbers is a simple multiplication. Therefore, all of the entries in the mathematical table are 1, because all that we are doing is multiplying 1 by 1. This process may seem to be trivial, but what we have done is come up with a mathematical mimic or *mathematical representation* of our soldier group. The two tables, side by side, are completely analogous, since the results of the multiplication in the mathematical representation are always the same as the number that we assigned to that particular operation. This is hardly surprising since we gave all the operations the number 1. However, we have discovered a very important point, which is: it is possible to mimic a group in terms of simple arithmetic. Indeed, it is also possible to say that assigning all of the operations the number 1 is always going to be a representation of any group we can dream up.* Of course, this is a lousy representation, because a lot of the information is lost. This representation is sometimes called an *unfaithful representation*.

Can you come up with other sets of numbers which are accurate mathematical representations of our soldier group? Remember the original multiplication table and the mathematical multiplication table must be completely analogous. For example, if you assign the value -1 to the R operation, whenever it appears in the original table, it must also appear in the mathematical table.

Answer

There is one trivial set, which is:

$E = 0, R = 0, L = 0, A = 0$, which tells us no more that when all elements are equal to 1.

There is only one other possible answer using real numbers, which is the following:-

$E = 1, R = -1, L = -1, A = 1$

The multiplication tables are given below for this non-trivial set.

	E	R	L	A
E	E	R	L	A
R	R	A	E	L
L	L	E	A	R
A	A	L	R	E

	1	-1	-1	1
1	1	-1	-1	1
-1	-1	1	1	-1
-1	-1	1	1	-1
1	1	-1	-1	1

From this we see that the multiplication tables are completely analogous. We have assigned a value of -1 to the operation R. Whenever R appears in the original table, -1 appears in the mathematical table. Similarly, we have assigned a value of 1 to A, and whenever A appears in the original table, a 1 appears in the mathematical table. We now have two non-trivial mathematical representations of our soldier group. These representations can be expressed in tabular form as follows:

---

* If you are in the position of having already completed your group theory course, take a look at the character tables to prove this point.

	E	R	L	A
Rep 1	1	1	1	1
Rep 2	1	−1	−1	1

Where Rep 1 = representation 1, and Rep 2 = representation 2.

So, let us recap on where we have got to. We have shown that a mathematical group can be defined by a set of rules. We have also seen that there are many different examples of groups. One example is the soldier on parade, where the commands which the soldier follows form a mathematical group. We can take this one stage further and assign all of the group operations a number. Certain numbers are mathematical representations of the group operations insofar as they mimic the results of group operations. For example, the command 'turn right', which is given the symbol R is replaced in the mathematical table by 'multiply by −1' or 'multiply by 1', depending on which representation we use. Notice that we have replaced the *commands* or the *operations* with a mathematical representation. An important point is that *only certain mathematical representations actually mimic the group*, we cannot use any old combination of numbers; they simply would not mimic the group accurately. Overall, we have now completed the transformation of a set of commands to a soldier on parade into a completely mathematical setting. This is an important step to make in understanding the use of group theory.

3.5 Representations of molecular symmetry

Since the symmetry operations of molecules also form a mathematical group, it should be possible to come up with mathematical representations of the symmetry operations. Let us try this with the symmetry operations of the point group C_{2v}. The symmetry operations of the group are E, C_2, $\sigma_v(xz)$ and $\sigma_v'(yz)$.

Draw up the multiplication table for the C_{2v} symmetry operations. When two answers are possible for the same combination, e.g. $C_2\sigma_v = E$ or σ_v', put both possible answers in the table. It might help to picture the H_2O molecule here. Remember that σ_v corresponds to reflection in the xz plane.

Answer

	E	C_2	σ_v	σ_v'
E	E or σ_v'	C_2 or σ_v	C_2 or σ_v	E or σ_v'
C_2	C_2 or σ_v	E or σ_v'	E or σ_v'	C_2 or σ_v
σ_v	C_2 or σ_v	E or σ_v'	E or σ_v'	C_2 or σ_v
σ_v'	E or σ_v'	C_2 or σ_v	C_2 or σ_v	E or σ_v'

This gives us a rather complicated table, where there is a variety of possible combinations. Is it possible to come up with a mathematical representation of the multiplication table above, where the answer to the combination of operations gives the mathematical representation of at least one of the answers?

Can you come up with mathematical representations of the multiplication table above, using real numbers? (Ignore the trivial answers.) The best method here is to draw up the multiplication tables for your trial combinations alongside the table given above.

$E = 1 \quad C_2 = 1 \quad \sigma_v = 1 \quad \sigma_v' = 1$

Answer

The following (non-trivial) answers are possible:-

$E = 1, C_2 = 1, \sigma_v = 1, \sigma_v' = 1$
$E = 1, C_2 = 1, \sigma_v = -1, \sigma_v' = -1$
$E = 1, C_2 = -1, \sigma_v = 1, \sigma_v' = -1$
$E = 1, C_2 = -1, \sigma_v = -1, \sigma_v' = 1$

It is quite easy to check that the mathematical representations mimic the C_{2v} operations by drawing up the appropriate multiplication tables. Two of the mathematical tables are given below. Note that the tables on the left have only the symmetry operation which can be represented by that number.

	E	C_2	σ_v	σ_v'		1	-1	-1	1
E	E	C_2	σ_v	σ_v'	1	1	-1	-1	1
C_2	C_2	E	σ_v'	C_2	-1	-1	1	1	-1
σ_v	σ_v	σ_v'	E	σ_v	-1	-1	1	1	-1
σ_v'	σ_v'	C_2	σ_v	E	1	1	-1	-1	1

	E	C_2	σ_v	σ_v'		1	1	-1	-1
E	E	C_2	σ_v	σ_v'	1	1	1	-1	-1
C_2	C_2	E	σ_v'	σ_v	1	1	1	-1	-1
σ_v	σ_v	σ_v'	E	C_2	-1	-1	-1	1	1
σ_v'	σ_v'	σ_v	C_2	E	-1	-1	-1	1	1

Just like the soldier group, it is possible to represent the symmetry operations of a molecule with a mathematical operation. For example, we can replace the 'rotate by 180°' operation with 'multiply by 1' or 'multiply by –1' depending on the particular representation. The mathematical operations are completely analogous to the symmetry operations, and follow all the rules of a mathematical group. We conclude, therefore, that we can represent the symmetry of a molecule mathematically.

For the H_2O molecule, write the representations in a tabular form, as we did for the soldier on parade group on page 41.*

Answer

	E	C_2	σ_v	σ_v'
Rep 1	1	1	1	1
Rep 2	1	-1	1	-1
Rep 3	1	-1	-1	1
Rep 4	1	1	-1	-1

---

* For those who have already studied group theory, you will probably have noticed that we have generated the irreducible representations of the C_{2v} character table.

Notice that these are the simplest representations that we can come up with. None of the representations can be broken down into simpler representations, or expressed as the sum of other representations. These simple representations are known as *irreducible representations*. In other words, these representations cannot be broken down into anything simpler (without using complex numbers). All of the irreducible representations for each symmetry point group have been worked out previously. These representations are listed in tables known as *character tables*. The important character tables in chemistry are listed in Appendix 1 of this book; take a look at them now. You will notice that the representations appear in tabular form. Some of the representations may not make sense in the light of what we have covered in this section, such as those representations with 2s and 0s in them. These 'unusual' representations will be explained in the next section.

This brings us to the end of the section. We have covered some important principles of how a group can be defined and also represented mathematically. The representations, as we shall see in later sections, become very important in figuring out the possible energy levels of a molecule. However, to take the theory much further, we need to begin to use more advanced mathematics (although, even the more difficult mathematics is within the knowledge of most chemistry undergraduates). The next section covers the mathematics which we need to understand more fully the role of mathematical representations in describing the possible energy levels of a molecule. Whatever your background in mathematics, it is worth attempting the next section as many of the puzzles of group theory are addressed therein. But, if you really feel uncomfortable with the mathematics, it is possible to miss out the next section and still be able to complete the chemistry-related exercises in the following sections. You will simply have to accept the conclusions listed at the beginning of Section 5. If you do miss out the next section, at least read the conclusions at the end, to help you understand the terminology.

3.6   Summary

What have we learnt from this short section?

- A group can be defined mathematically, and there is a set of mathematical rules which defines a group.

- There are many examples of mathematical groups of which molecular symmetry is just one type.

- It is always possible to represent the *operations* of a group with numbers; the simplest form of which is to replace the operations with 1 or −1.

- The simplest representations cannot be expressed in terms of other representations, and are called *irreducible representations*.

- Only certain mathematical representations are true representations of a group; it is not possible to pick any old set of numbers.

More representations

More representations

4.1   Other representations and bases

In this section we will see how other representations can be built up using slightly more sophisticated mathematical techniques. Whereas it is true that those who are not mathematically minded may find this section hard work (and, as I have already said, can miss it out altogether if they want) it is well worth trying to understand the material in this section. The mathematics, whilst being more sophisticated, can be picked up fairly easily. In particular, the rules of matrix algebra are required. Most standard mathematics books have sections on matrices.

In the last section, we saw how the symmetry operations of a group could be represented mathematically by a series of numbers. These numbers are representations because they form a group analogous to the symmetry operations, but, can we come up with even more representations? Clearly, for the examples seen in the last chapter, it is impossible to imagine any other representations based on real numbers. To be able to come up with more representations we have to change the way we look at the object that we are operating on—be it a soldier or a molecule.

Let us turn back to our soldier on parade. We can define the soldier in a more mathematical way, by placing the soldier on the origin of a set of normal cartesian axes. Since we are only dealing with two dimensions, we need only to use two axes. The origin is defined as the centre of rotation of the soldier. We can now define the position of any general point with the coordinates x and y. Let us say that the coordinates refer to the tip of the soldier's rifle.

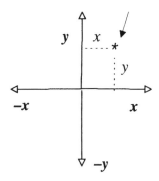

We have defined part of the soldier more mathematically. We now have several questions to answer. How do these coordinates behave under the operations? And can we come up with some way of representing the operation on these coordinates mathematically? If we can do this, does our new mathematical representation of the operations form a group, just like the representations that we have already seen in the previous section?

By sketching the axes and coordinates again, show what happens to the coordinates upon the *R* operation. Note that the axes themselves *do not undergo the R operation*, the axes act as a fixed frame of reference.

Answer

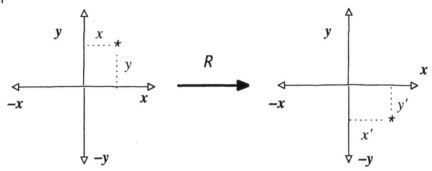

The effect of the rotation is to carry the point to a new position, defined by the coordinates *x*' and *y*'.

How can the new coordinates, *x*' and *y*', be expressed in terms of the old coordinates *x* and *y*?

Answer

The new *x* coordinate, *x*', has the value of the old *y* coordinate.

The new *y* coordinate, *y*', has the value of −(the old *x* coordinate).

Since the new coordinates can be expressed in terms of the old coordinates, we can write the following.

$$\begin{pmatrix} x \\ y \end{pmatrix} \xrightarrow{R} \begin{pmatrix} y \\ -x \end{pmatrix}$$

Notice that the coordinates are written in a 'column' format, with the *x* coordinate value at the top. We could have equally have written them in a 'row' format, like (*x y*), but we chose to write them in column format. We will use this format for coordinates from now on.

Therefore, it appears as if we can use the cartesian coordinates as a mathematical *basis* for our representation. The word 'basis' here is important; it is simply a mathematical way of describing the position of our soldier. We could equally have chosen something else to represent the soldier's position, for instance we could have chosen a function or a set of vectors. Both of these could have served as a basis for describing the soldier's direction. Whatever basis we choose, all it needs to do is describe the object that we are studying. For example—and we will see this later—a molecule can be described by the coordinates of its atoms, the positions of its bonds, the wavefunctions of the atoms.

By sketching the axes and coordinates show how the x and y coordinates change when the operations L, A or E are carried out on them. Write the result of the operation on the coordinates in a similar manner to the R operation above.

Answer

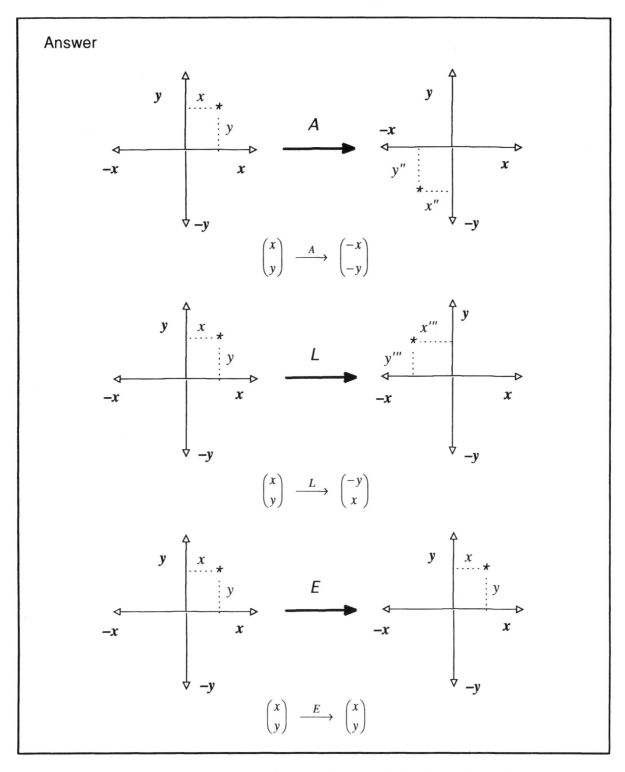

$$\begin{pmatrix} x \\ y \end{pmatrix} \xrightarrow{A} \begin{pmatrix} -x \\ -y \end{pmatrix}$$

$$\begin{pmatrix} x \\ y \end{pmatrix} \xrightarrow{L} \begin{pmatrix} -y \\ x \end{pmatrix}$$

$$\begin{pmatrix} x \\ y \end{pmatrix} \xrightarrow{E} \begin{pmatrix} x \\ y \end{pmatrix}$$

Now that we have a mathematical way of representing the position of part of our soldier, can we also represent the *operations* on the coordinates mathematically, in a similar manner to the previous section? In other

words, can we replace the $\xrightarrow{T}$ (where T is the operation) with a mathematical device to convert the coordinates as shown above? And, does this new mathematical representation of operations form a mathematical group in the same way? The answer to the first and second questions is yes, but, we cannot use simple integers as before, we need to turn to matrices. Matrices are a whole branch of mathematics in themselves, and the rules for manipulating matrices have been worked out. Help about matrices can be found in any good general mathematics book. I have assumed a basic knowledge of matrix algebra in what follows. The answer to the third question is also yes, and we will demonstrate this later.

4.2   Matrices as representations

If the x and y coordinates are represented as a column matrix, then we can represent the operations as individual matrices. If we also say that the act of the operation matrix operating on the coordinates matrix is represented by a simple multiplication of the two matrices, following all of the rules of matrix multiplication, then we are able to write a mathematical expression, which describes the operation of R on the two coordinates. Notice that the R matrix operates *on* the coordinate matrix, therefore, we choose to *premultiply* the coordinate matrix with the operator matrix. (Remember that matrix multiplication is non-commutative.) Thus, in terms of matrix algebra, the R operation is represented as follows:

$$\text{Let } R = \begin{pmatrix} 0 & 1 \\ -1 & 0 \end{pmatrix}$$

$$\begin{pmatrix} x \\ y \end{pmatrix} \xrightarrow{R} \begin{pmatrix} y \\ -x \end{pmatrix} \quad \text{is represented as } R\begin{pmatrix} x \\ y \end{pmatrix}, \text{ where } R \text{ is the operation matrix.}$$

Thus:

$$R\begin{pmatrix} x \\ y \end{pmatrix} = \begin{pmatrix} 0 & 1 \\ -1 & 0 \end{pmatrix}\begin{pmatrix} x \\ y \end{pmatrix} = \begin{pmatrix} y \\ -x \end{pmatrix}$$

The matrix which represents R performs the necessary transformation of the x and y coordinates, following all the rules of matrix multiplication. Notice that the operation matrix is a square 2×2 matrix in this case.

Write down the 2×2 matrices which represent the L, A and E operations.

Answer

$$L = \begin{pmatrix} 0 & -1 \\ 1 & 0 \end{pmatrix} \qquad A = \begin{pmatrix} -1 & 0 \\ 0 & -1 \end{pmatrix} \qquad E = \begin{pmatrix} 1 & 0 \\ 0 & 1 \end{pmatrix}$$

So, we have answered the first and second questions, that we posed earlier. It is possible, with matrices, to represent mathematically the operations of R, L, A and E on the two basis coordinates. The third question is: do these matrices form a mathematical group in their own right? We need to look back at our definition of a group in Section 3.3 to answer this. If our matrices form a mathematical group, then we can say that the group is a mathematical representation of the group of operations. The rules of a mathematical group are:

Rule 1   The combination of any two of the group elements must be the same as a single group element.

We need to test whether this rule is obeyed by our matrices. For example, the combination *RL* is represented by this matrix expression:

$$RL = \begin{pmatrix} 0 & 1 \\ -1 & 0 \end{pmatrix} \begin{pmatrix} 0 & -1 \\ 1 & 0 \end{pmatrix} = \begin{pmatrix} 1 & 0 \\ 0 & 1 \end{pmatrix} = E$$

This shows that the combination of *R* with *L* in terms of their matrix representations gives the same matrix which is identified with the operation *E*. This, of course, correlates with our definition of a group, and shows that this particular multiplication of matrices gives another group element.

Write down and calculate the matrix representations for the following combinations: *LR, AR, LA*. Satisfy yourself that the matrix answers to these combinations are members of the representative matrix group.

Answer

$$LR = \begin{pmatrix} 0 & -1 \\ 1 & 0 \end{pmatrix} \begin{pmatrix} 0 & 1 \\ -1 & 0 \end{pmatrix} = \begin{pmatrix} 1 & 0 \\ 0 & 1 \end{pmatrix} = E$$

$$AR = \begin{pmatrix} -1 & 0 \\ 0 & -1 \end{pmatrix} \begin{pmatrix} 0 & 1 \\ -1 & 0 \end{pmatrix} = \begin{pmatrix} 0 & -1 \\ 1 & 0 \end{pmatrix} = L$$

$$LA = \begin{pmatrix} 0 & -1 \\ 1 & 0 \end{pmatrix} \begin{pmatrix} -1 & 0 \\ 0 & -1 \end{pmatrix} = \begin{pmatrix} 0 & 1 \\ -1 & 0 \end{pmatrix} = R$$

The combinations all show that multiplication of the representative matrices gives another matrix which is a representative of another of the group operations. This is, in fact, a general result for this group.

Rule 2   One element in the group must do nothing—the identity element.

This, of course, is the identity matrix: $\begin{pmatrix} 1 & 0 \\ 0 & 1 \end{pmatrix}$

Rule 3   The combinations of group elements must be associative. In other words, this general case must be true:

$$A(BC) = (AB)C$$

This is a general property of matrix algebra, and will not be proved here.

Rule 4   Every group element must have an inverse (i.e. something to undo what the operation does) which is also an element of the group.

We need to show that any matrix in the group can be combined with another matrix in the group to give the 'nothing overall has happened' result, in other words, gives the identity matrix.

Write down the following combinations in matrix form and show that they give the identity matrix.

RL, LR, AA, EE.

Answer

$$LR = \begin{pmatrix} 0 & -1 \\ 1 & 0 \end{pmatrix}\begin{pmatrix} 0 & 1 \\ -1 & 0 \end{pmatrix} = \begin{pmatrix} 1 & 0 \\ 0 & 1 \end{pmatrix} = E$$

$$RL = \begin{pmatrix} 0 & 1 \\ -1 & 0 \end{pmatrix}\begin{pmatrix} 0 & -1 \\ 1 & 0 \end{pmatrix} = \begin{pmatrix} 1 & 0 \\ 0 & 1 \end{pmatrix} = E$$

$$AA = \begin{pmatrix} -1 & 0 \\ 0 & -1 \end{pmatrix}\begin{pmatrix} -1 & 0 \\ 0 & -1 \end{pmatrix} = \begin{pmatrix} 1 & 0 \\ 0 & 1 \end{pmatrix} = E$$

$$EE = \begin{pmatrix} 1 & 0 \\ 0 & 1 \end{pmatrix}\begin{pmatrix} 1 & 0 \\ 0 & 1 \end{pmatrix} = \begin{pmatrix} 1 & 0 \\ 0 & 1 \end{pmatrix} = E$$

Therefore, our matrices form a group by themselves, and, this matrix group behaves in an identical manner to the real operations of R, L, A and E. Thus, the matrices are, by themselves, a *mathematical representation* of our soldier group. So, with the results from the previous section we now have three different representations of the group, which can be expressed in tabular form, see below. We will see later, that we can write similar tables for molecular symmetry groups. These tables are a useful means of collecting together mathematical representations.

	E	R	L	A
Rep 1	1	1	1	1
Rep 2	1	-1	-1	1
Rep 3	$\begin{pmatrix} 1 & 0 \\ 0 & 1 \end{pmatrix}$	$\begin{pmatrix} 0 & 1 \\ -1 & 0 \end{pmatrix}$	$\begin{pmatrix} 0 & -1 \\ 1 & 0 \end{pmatrix}$	$\begin{pmatrix} -1 & 0 \\ 0 & -1 \end{pmatrix}$

Where Rep = representation.

In fact, the use of matrices as mathematical representations of molecular symmetry groups is a key part of group theory. Can we go further? Can we come up with similar matrix representations of molecular symmetry groups? And, if we can, what does this tell us? The next step then is to see if we can indeed come up with representations of molecular symmetry operations, in a similar manner to those above.

Where does all of this lead? This question will hopefully be answered at the beginning of Section 5. We will then see that such representations become an extremely powerful means of examining the properties of a molecule.

4.3   Matrix representations of molecular symmetry

First, we need to pick a property of the molecule that we wish to study. For example, we might be interested in the way atoms move relative to one another (this will eventually lead us into studying molecular vibrations), or we might be interested in how to combine atomic orbitals together within the shape of the molecule. For any of these properties we need to express it mathematically. Therefore, we could represent the motions of atoms with respect to one another using displacement coordinates,* or we could represent the atomic orbitals of the atoms with the mathematical function which describes the particular orbitals. This mathematical description of the molecule becomes our *basis* for the representation of the symmetry operations. Hopefully, this will all be made clear with the following example.

Second, we need to find the matrices—as we did above for the soldier group—which describe the transformation of our chosen basis under the symmetry operations of the point group in which the molecule lies. These matrices should form a mathematical representation of the group. Again, this can be made clearer with an example.

Let us start by considering the water molecule.

Without looking back, what are the symmetry elements in the water molecule? What point group is assigned to it?

Answer

These are E, C_2, σ_v and σ_v'. Point group assignment: C_{2v}

---

* A displacement coordinate is the 'motion' of an atom expressed in terms of cartesian coordinates. For example if an atom in position x on the x axis moves to a new position $x + \delta x$, then the displacement coordinate is simply δx.

Say that we are interested in the way atoms move with respect to one another, with the eventual aim of studying the vibrations of the molecule. (After all, the vibration of a molecule can always be represented by the relative motions of the atoms in the molecule.) We need to chose a basis to represent this. We can view the position of each atom as being defined by a single set of cartesian axes, say with its origin on the oxygen atom. We can then define how each of these atoms 'moves' with respect to the axes by the change in the x, y and z values of their individual coordinates, i.e. δx, δy and δz, which are called *displacement coordinates*. The displacement coordinates of each atom can be depicted by three orthogonal (i.e. at 90° to one another) arrows (see diagram). Notice that we have used a right hand set of cartesian axes to depict the displacement coordinates; this shows the x axis coming 'out' of the plane of the paper.

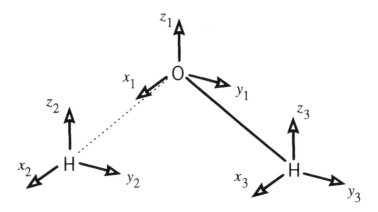

In the figure above the displacement coordinates on each atom are represented by x_a, y_a and z_a, where a is the 'number' of the atom. The δ has been dropped for convenience. We can now use the displacement coordinates on each atom as a basis for our mathematical representation for the symmetry operations of the molecule.

In an analogous way to the soldier example, what happens to these basis displacement coordinates when a symmetry operation is performed upon them? The example below shows what happens when C_2 is performed.

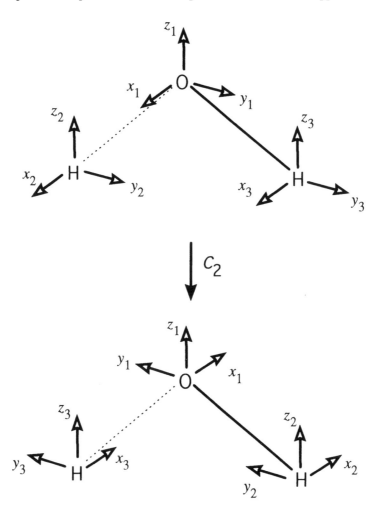

Therefore, in an analogous way, we can write a matrix which represents this transformation on all nine of the displacement coordinates.

$$
\begin{pmatrix}
-1 & 0 & 0 & 0 & 0 & 0 & 0 & 0 & 0 \\
0 & -1 & 0 & 0 & 0 & 0 & 0 & 0 & 0 \\
0 & 0 & 1 & 0 & 0 & 0 & 0 & 0 & 0 \\
0 & 0 & 0 & 0 & 0 & 0 & -1 & 0 & 0 \\
0 & 0 & 0 & 0 & 0 & 0 & 0 & -1 & 0 \\
0 & 0 & 0 & 0 & 0 & 0 & 0 & 0 & 1 \\
0 & 0 & -1 & 0 & 0 & 0 & 0 & 0 & 0 \\
0 & 0 & 0 & -1 & 0 & 0 & 0 & 0 & 0 \\
0 & 0 & 0 & 0 & 0 & 1 & 0 & 0 & 0
\end{pmatrix}
\begin{pmatrix}
x_1 \\ y_1 \\ z_1 \\ x_2 \\ y_2 \\ z_2 \\ x_3 \\ y_3 \\ z_3
\end{pmatrix}
=
\begin{pmatrix}
-x_1 \\ -y_1 \\ z_1 \\ -x_3 \\ -y_3 \\ z_3 \\ -x_2 \\ -y_2 \\ z_2
\end{pmatrix}
$$

This gives us a rather cumbersome 9×9 matrix, which represents the C_2 operation on our basis set of vectors. We can perform the same exercise for the action of the σ_v reflection plane to come up with a similar 9×9 matrix which mimics the operation of the plane.

By sketching the basis set of vectors above and showing the transformation of these vectors upon the $\sigma_v(xz)$ reflection plane, come up with a 9×9 matrix to represent the σ_v reflection plane operation.

Answer

The reflection plane gives the following result.

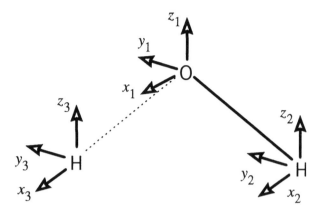

Which is represented by the following matrix multiplication:

$$
\begin{pmatrix}
1 & 0 & 0 & 0 & 0 & 0 & 0 & 0 & 0 \\
0 & -1 & 0 & 0 & 0 & 0 & 0 & 0 & 0 \\
0 & 0 & 1 & 0 & 0 & 0 & 0 & 0 & 0 \\
0 & 0 & 0 & 0 & 0 & 0 & 1 & 0 & 0 \\
0 & 0 & 0 & 0 & 0 & 0 & 0 & -1 & 0 \\
0 & 0 & 0 & 0 & 0 & 0 & 0 & 0 & 1 \\
0 & 0 & 0 & 1 & 0 & 0 & 0 & 0 & 0 \\
0 & 0 & 0 & 0 & -1 & 0 & 0 & 0 & 0 \\
0 & 0 & 0 & 0 & 0 & 1 & 0 & 0 & 0
\end{pmatrix}
\begin{pmatrix}
x_1 \\ y_1 \\ z_1 \\ x_2 \\ y_2 \\ z_2 \\ x_3 \\ y_3 \\ z_3
\end{pmatrix}
=
\begin{pmatrix}
x_1 \\ -y_1 \\ z_1 \\ x_3 \\ -y_3 \\ z_3 \\ x_2 \\ -y_2 \\ z_2
\end{pmatrix}
$$

Without taking the point further, it is also possible to come up with 9×9 matrices which represent the other symmetry operations of the point group. It is also possible to show that these matrices form a representation of the C_{2v} point group. The problem with them, of course, is that they are very cumbersome to use, and certainly not easy to write out accurately.

Before we go on to see how to simplify the matrices, it is necessary that we look at another example of this type of basis set, to show some of the possible complications that can arise. In the above example, the displacement coordinates after transformation could be described fully in terms of one of the original displacement coordinates. However, there are cases where this is not always true, sometimes the transformed displacement coordinate needs to be described as a combination of two or more of the original displacement coordinates: this complicates matters slightly. The following example illustrates this point.

Consider the ammonia molecule. Let us say we are interested in the how the nitrogen atom moves in the xy plane (ignore the hydrogen atoms and the N atom z displacement coordinate to make the example simpler). To study this we create a set of two displacement coordinates on the nitrogen atom as in the figure.

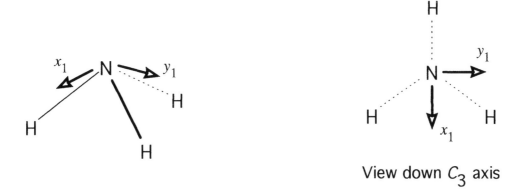

View down C_3 axis

The ammonia molecule is in the C_{3v} point group, with the following symmetry elements: E, $2C_3$ and $3 \times \sigma_v$. Let us see the effect of the C_3 operation on our basis set of displacement coordinates. The best way to view the operation is down the C_3 axis.

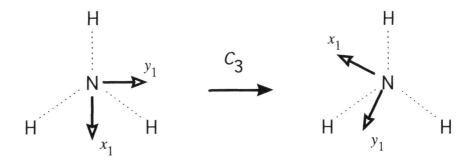

How can we now describe the new displacement coordinates in terms of the old displacement coordinates? Clearly, in this case there is no longer a 1:1 relationship between the original displacement coordinates and the new displacement coordinates. To answer the question, we must describe the new displacement coordinates as a *linear combination* of the original displacement coordinates. To do this we need to resort to some trigonometry. We are treating the displacement coordinate as vectors in this sense. What we are actually doing is regarding the displacement coordinates as the fractional components of the cartesian axes which define them. Since the unit cartesian axes have magnitude and direction we can treat the displacement coordinates as vectors for this simple addition. The figure below shows the new x_1 displacement coordinate, x_1', mapped onto the old displacement coordinates, shown as x_1 and y_1 and $-x_1$ and $-y_1$.

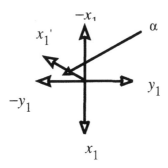

The new x_1' displacement coordinate lies between the $-x_1$ and $-y_1$ displacement coordinates, and, using some simple trigonometry, we can describe the x_1' displacement coordinate in terms of a combination of the $-x_1$ and $-y_1$ displacement coordinates. For x_1', this gives:

$$x_1' = -x_1(\sin \alpha) + -y_1 (\cos \alpha)$$

where α is the angle between the $-y_1$ and x_1' displacement coordinates in the figure. This shows that the new displacement coordinate can be expressed as a combination of the old displacement coordinates.

More generally, this gives (and we now use θ (= 90 + α), which is the angle through which the symmetry operation rotates, and also that $\sin \alpha = -\cos(90 + \alpha) = -\cos \theta$ and $\cos \alpha = \sin (90 + \alpha) = \sin \theta$):

$$x_1' = x_1(\cos \theta) - y_1(\sin \theta)$$
$$y_1' = x_1(\sin \theta) + y_1(\cos \theta)$$

We can now represent this in a matrix format:

$$\begin{pmatrix} \cos \theta & -\sin \theta \\ \sin \theta & \cos \theta \end{pmatrix} \begin{pmatrix} x_1 \\ y_1 \end{pmatrix} = \begin{pmatrix} x_1' \\ y_1' \end{pmatrix}$$

This turns out to be a general result in group theory. We will use this result in a later section. For the interested reader, the derivation of these general cases can be found in more advanced books on group theory, but since their derivation is not important for solving simple problems in chemical group theory, we will not examine them further. The most important message is that we can generate matrix representations using displacement coordinates on each atom as a basis. Unfortunately, these matrices tend to be large and cumbersome.

Before we move on to how these matrix representations can be made less cumbersome, it is worth looking at one more basis set to describe a molecule. (In sections 6 onwards we will look at different basis sets to describe molecules. These other basis sets will be introduced then, but they follow exactly the same principles as the basis sets described in this chapter.) This basis set is the atomic orbitals of the individual atoms which are part of a molecule.

Again, consider the water molecule. In particular, consider the valence atomic orbitals on the oxygen atom. These are $2s$, $2p_x$, $2p_y$ and $2p_z$. Let us see how each one of these orbitals behaves upon the symmetry operations of the C_{2v} point group in which the symmetry operations of the water molecule lie. For the moment we will ignore the hydrogen atom $1s$ orbitals

We shall first consider the $2s$ orbital. Without sketching the orbital itself, which is completely spherical it is easy to imagine that all of the symmetry operations leave the s orbital completely unchanged.

Using numbers, think of the simplest way you could mathematically represent the way the $2s$ orbital on the oxygen atom is (or is not!) changed under the symmetry operations of the C_{2v} point group. Draw up your simple representation in tabular form, as shown at the end of the last section.

Answer

Since the s orbital does not change under any of the symmetry operations, we can represent it simply as follows:

	E	C_2	σ_v	σ_v'
	1	1	1	1

Of course, we have seen this representation before on page 43, and we know that it is an accurate representation of the group. This representation is given another name, which is the *totally symmetric* representation.

Now, come up with mathematical representations for the $2p$ orbitals of the O atom in H_2O. It will help to sketch the orbitals as they are transformed under the symmetry operations of the group. If an orbital is reversed, then the symmetry operation can be given the value -1.

p_y

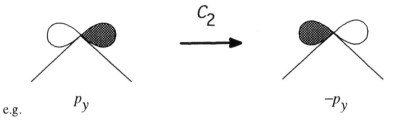

e.g. p_y $-p_y$

Therefore, C_2 takes the value -1 in our representation for the p_y orbital.

p_x

p_z

Answer

	E	C_2	σ_v	σ_v'
s, p_z	1	1	1	1
p_x	1	-1	1	-1
p_y	1	-1	-1	1

The table shows that the representations are actually very simple. Indeed, we have encountered all of these representations in the previous chapter, and we know that they are accurate mathematical representations of the group. We will cover this again, when we come to studying character tables later in this section, but, it is worth noting here that the representations of the atomic orbitals of the atom which lies on the point of the group (i.e. is unshifted in position upon the symmetry operations of the group) are very simple indeed.

How about the orbitals of the hydrogen atoms? The effect of the symmetry operation is shown in the figure below.

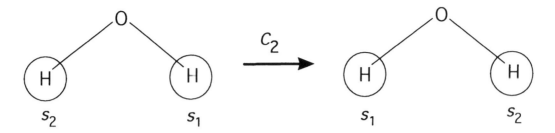

The s orbitals are mathematical functions and we can, as before, represent this in matrix notation, as follows:

$$(s_1\ s_2)\begin{pmatrix}0 & 1\\ 1 & 0\end{pmatrix} = (s_2\ s_1)$$

You might, at this point, be asking why the matrix which represents the functions has been written as a row matrix, and the operation matrix has been written *after* the function matrix. It is to do with the different ways in which coordinates and functions transform under symmetry operations. The matrix to transform coordinates under a symmetry operation can also be used to transform a column matrix of functions. This is all apparently OK. However, what goes unnoticed is a transformation of the cartesian axes which are used to describe the functions and the vectors. This transformation of axes is different for vectors than for functions. This axes transformation does become apparent if group theory is used at a more advanced level. By writing the function matrix as a row matrix we avoid this problem. Therefore, despite the fact that it makes no difference to this text, it is important to establish the correct means of representing the transformation of functions. In any case, the matrix used to transform the row matrix of functions forms an accurate representation of the group. If this makes little sense to you, it is probably not worth spending time worrying about the reasons for it, as when we come to using group theory at this level, it is not important.

Write down the matrices which represent the transformations of the two hydrogen $1s$ orbitals of H_2O under the symmetry operations of the C_{2v} point group.

Answer

$$E = \begin{pmatrix}1 & 0\\ 0 & 1\end{pmatrix} \qquad C_2 = \begin{pmatrix}0 & 1\\ 1 & 0\end{pmatrix}$$

$$\sigma_v = \begin{pmatrix}0 & 1\\ 1 & 0\end{pmatrix} \qquad \sigma_v' = \begin{pmatrix}1 & 0\\ 0 & 1\end{pmatrix}$$

These matrices now form a representation of the hydrogen $1s$ orbitals under the symmetry operations of the point group, and can be expressed in tabular form as follows:

	E	C_2	σ_v	σ_v
$1s$ on H	$\begin{pmatrix} 1 & 0 \\ 0 & 1 \end{pmatrix}$	$\begin{pmatrix} 0 & 1 \\ 1 & 0 \end{pmatrix}$	$\begin{pmatrix} 0 & 1 \\ 1 & 0 \end{pmatrix}$	$\begin{pmatrix} 1 & 0 \\ 0 & 1 \end{pmatrix}$

So, in this part we have seen that matrices can be used to represent the symmetry operations of a molecule. The matrices themselves form a mathematical group which behaves identically to the symmetry operations of the point group (this is more technically known as an isomorphism between the matrices and the symmetry operations).

To derive these matrices we need to describe a mathematical basis on which the matrices can operate. The mathematical basis can be anything that we might be interested in, for example, the wavefunctions of the orbitals, the displacement coordinates of the atoms. In fact, we are only limited by our imagination in the basis we choose to describe the molecule that we are interested in.

The problem with the matrices is that they can be very cumbersome to use, and do not really lend themselves to the easy analysis of a molecule. Moreover, what do the matrices really mean? In the next part, we will see that we can make the matrices much more 'user-friendly'. We will also see that we can use the fact that the matrices form a mathematical group to simplify them even further.

4.4   Reducible representations and irreducible representations

In this part we will see how we can turn the fact that the matrices form a mathematical group to our advantage. We will make use of a few very powerful properties of mathematical groups. We will simply use these mathematical properties without proof or derivation; for the interested reader the proofs can be found in any advanced textbook on chemical group theory.

We will start this part with an example.

Without looking back, determine the point group of the ammonia molecule?

Answer

C_{3v}

Consider the ammonia molecule again. Set up three-dimensional displacement coordinates (x_1, y_1 and z_1) on the nitrogen atom. (This is an extension to what we did above with two dimensional displacement coordinates.) Come up with the 3×3 matrix which acts as a representation of this basis under the $C_3{}^1$ symmetry operation of the C_{3v} point group. (See pages 57 and 58.)

Answer

$$C_3^1 = \begin{pmatrix} \cos\theta & -\sin\theta & 0 \\ \sin\theta & \cos\theta & 0 \\ 0 & 0 & 1 \end{pmatrix}$$

This 3×3 matrix has some interesting features. First, it is a simple extension of the 2×2 case that we encountered above for the x and y displacement coordinates. Second, it essentially contains two matrices, there is the 2×2 matrix which represents the x and y displacement coordinates and there is the 1×1 matrix (i.e. the number 1) which represents the z displacement coordinate, this is depicted schematically below.

Therefore, we can envisage that the 3×3 matrix which represents the transformation of the three displacement coordinates is made-up of two smaller matrices. In essence, we have *reduced* the 3×3 matrix to two smaller matrices. Obviously, we can only go so far with this *reduction* procedure, until we get to a point where the resulting matrices are no longer reducible, in other words they become *irreducible*.

Tne example given above is by no means a proof of what I am about to state, but it does illustrate the point that we can reduce some matrix representations into 'smaller' matrix representations. The mathematics involved in doing this reduction is beyond this text, as in many cases the conversion of a matrix representation into the 'block diagonal' form seen in this example is not simple. (In the example above the 3×3 matrix representation appears immediately and conveniently in its block diagonal form, which can be 'reduced' by inspection.) What follows is something that comes right out of mathematical group theory, and it has an important place in chemical group theory.

For any matrix representation of the symmetry operations of a point group, this representation can be reduced into 'smaller' irreducible representations.

The matrix representation that can be reduced is called a *reducible representation*. The 'smaller' representations into which the reducible representations are reduced are called *irreducible representations*.

In other words, we can come up with any matrix representation of molecular symmetry operations, and this representation can *always* be expressed in terms of *irreducible representations* of that particular point group. In the next section we will see how this reduction is done in practice. We will also see that all of the irreducible representations have been evaluated for all of the chemical point groups.

4.5   Characters

We have one more problem to overcome which is the difficulty of handling large matrices. Can we simplify the large matrices to make them easier to handle? Well, there is one very important result from mathematical group theory which allows us to do this simplification.

Take a general square matrix. We can define a simple property of the matrix which is called the *trace* of the matrix. The trace of a matrix is the numerical addition of its diagonal elements. So for the general 4×4 matrix:

$$\begin{pmatrix} a & b & c & d \\ e & f & g & h \\ i & j & k & l \\ m & n & o & p \end{pmatrix}$$

the trace is simply the numerical value of $a + f + k + p$.

Calculate the traces of the following matrices:

$$\begin{pmatrix} 1 & 2 \\ 3 & 4 \end{pmatrix} \qquad \begin{pmatrix} 2 & 4 & 6 \\ 8 & 6 & 4 \\ 1 & 3 & 5 \end{pmatrix} \qquad \begin{pmatrix} 1 & 0 & 1 & -1 \\ 2 & -1 & 0 & 0 \\ 0 & 0 & 2 & -1 \\ -1 & 1 & 1 & 1 \end{pmatrix}$$

Answer

5, 13, 3

In chemical group theory, it turns out that the trace of a representation matrix is *characteristic* of its behaviour as a representation of a symmetry operation. In fact, one very important property of matrices in chemical group theory is that representation matrices do not change the value of their traces under all of the mathematical procedures involved in chemical group theory. Because the trace of a matrix is characteristic of the matrix in this way, the trace is also called the *character* of the matrix.

This property *greatly* simplifies our use of matrices in chemical group theory. All that we have to do is to figure out the trace of our representation matrices, and we need not be concerned with writing out the full matrix.

Going back to the reducible representation that we obtained for the hydrogen $1s$ orbitals in water, we obtained the following result:

	E	C_2	σ_v	σ_v'
$1s$ on H	$\begin{pmatrix} 1 & 0 \\ 0 & 1 \end{pmatrix}$	$\begin{pmatrix} 0 & 1 \\ 1 & 0 \end{pmatrix}$	$\begin{pmatrix} 0 & 1 \\ 1 & 0 \end{pmatrix}$	$\begin{pmatrix} 1 & 0 \\ 0 & 1 \end{pmatrix}$

This can now be conveniently rewritten in 'character' form as follows:-

	E	C_2	σ_v	σ_v'
$1s$ on H	2	0	0	2

This form is much easier to handle. Notice that only s orbitals that are *unshifted* by the symmetry operation contribute to the value of the character. For example, both s orbitals are left unshifted by the E symmetry operation, therefore, the character of the representation is 2 here. Whereas, the C_2 operation leaves neither of the two s orbitals unshifted, therefore, the character of the representation is 0 here. (The reason for this can be seen if we consider the form of the representation matrix. Any s orbital shifted from its original position *must* have a 0 for the diagonal element of the matrix on that matrix row, otherwise it contributes a 1 to the trace.)

The above shows how easy it is to come up with a reducible representation for the s orbitals in a molecule. We simply figure out the number of atoms that we are interested in upon each symmetry operation and for each unshifted atom add 1 to the character. We will practise this exercise later.

Come-up with a reducible representation in its character form for the *s* orbitals of the H atoms in NH_3. (Hint: first figure out all of the possible symmetry operations of the molecule and figure out the number of unshifted atoms on each symmetry operation. Then multiply the contribution of each unshifted H atom to the trace (which is always 1) by the number of unshifted H atoms.) Draw up your answer in tabular form as above.

Answer

	E	$C_3{}^1$	$C_3{}^2$	σ_v	$\sigma_v{}'$	$\sigma_v{}''$
No of unshifted H atoms	3	0	0	1	1	1
Contribution to trace per unshifted H atom	1	1	1	1	1	1
Reducible representation for *s* orbitals	3	0	0	1	1	1

A similar strategy can be applied to working out reducible representations for *p* orbitals. Again, we figure out the number of unshifted atoms upon the symmetry operation. This differs from the *s* orbital, the contribution of the three *p* orbitals on each atom to the trace of the matrix is *not* 1 in each case, but is given by the following equations:

Contributions to trace of matrix per unshifted atom using p *orbitals as a basis*

$$E = 3 \qquad\qquad i = -3$$
$$\sigma = 1 \qquad\qquad C_n = 1 + 2\cos(360/n)° \qquad\qquad S_n = -1 + 2\cos(360/n)°$$

These equations can be easily verified by writing out the appropriate full matrix which represents the transformation of *p* orbitals of the nitrogen atom in ammonia (see the top of page 62). Again, we will practise the use of these equations later.

Come-up with a reducible representation in its 'character form' for the *p* orbitals of the N atom in NH_3. (Hint: in this case the nitrogen atom is unshifted under all of the symmetry operations.)

Answer	E	$C_3{}^1$	$C_3{}^2$	σ_v	$\sigma_v{}'$	$\sigma_v{}''$
No of unshifted N atoms	1	1	1	1	1	1
Contribution to trace per unshifted N atom	3	0	0	1	1	1
Reducible representation for N p orbitals	3	0	0	1	1	1

It is easy to verify the equations for the contributions to the trace of the symmetry operation matrices, by deriving the whole matrix itself. We did a similar exercise earlier for the displacement coordinates on the nitrogen atom for ammonia. In fact, using three orthogonal displacement coordinates on each atom as a basis set allows us to use the same equations for the contribution to the trace of a representation matrix, as those for the p orbitals. We will practise this in Section 6.

4.6   Summary

We have seen in this section that it is possible to come up with an almost endless variety of mathematical representations of molecular symmetry operations. These representations are derived from describing the molecule with a mathematical basis, e.g. coordinates, functions, and then using matrices to represent the action of symmetry operations on this basis. The use of matrices is made a lot easier by considering only the traces (or characters) of the matrices. This removes the need to write out the whole of a representation matrix, which may be very large; it also makes the problem of coming up with a representation much easier, as we only need to figure out the number of unshifted atoms and the contribution of that particular basis set to the character of the transformation matrix.

We have also seen that a representation can be expressed as a combination of certain *irreducible* representations, and that all representations are ultimately irreducible representations or combinations of these irreducible representations.

Although we have only covered the mathematical basics of group theory, the concepts met in this section should prove useful in helping to understand what follows, which is mainly to do with using group theory in chemistry.

SECTION 5

The heart of group theory

The heart of group theory

5.1 What the mathematics told us

In Section 3, we took the first tentative steps towards putting our notion of symmetry on a mathematical basis. In Section 4, the mathematical basis was explained in more detail. In particular, we showed in Section 4 that we can chose a particular mathematical property of a molecule (e.g. the coordinates of its atoms, the wavefunctions of its atoms) and use this to come up with other representations of the symmetry operations of the molecule. In doing this we established a mathematical connection between the molecule and its symmetry.

Whether you managed to complete section 4 or not, the following points are crucial for the successful understanding and use of group theory in chemistry. If you did not manage to read section 4, then you will have to accept much of the following at face value.

1. We can represent a molecule in a mathematical way, e.g. with the coordinates of its atoms. This mathematical description of the molecule forms a *basis* for the symmetry operations.

2. Using this basis, we can generate mathematical representations of the symmetry operations with some simple rules. We will practice these rules in later sections.

3. The mathematical representations are either *reducible* or *irreducible*. Every reducible representation can be expressed as a combination of irreducible representations. We will see how to do this 'reduction' of a reducible representation later in this section.

4. The representations can be expressed very simply using numbers called *characters*.

5. The irreducible representations for all of the common point groups in chemistry have been worked out. The representations are tabulated in *character tables*. Character tables are explained in more detail below.

This is a condensation of the previous section. We can now progress to the more practical aspects of group theory in chemistry. We can also examine briefly why group theory is so useful in analysing molecules.

5.2 Character tables, irreducible representations and Mulliken symbols

We have now reached the point where we can describe the use of a character table in chemical group theory. So far we have seen that each molecule can be classified according to its symmetry operations into a point group. The point group of a molecule is shorthand for a collection of symmetry operations which can be carried out on the molecule. Some of the symmetry operations of the group 'behave' in a similar fashion and can be placed into classes. (In fact, if you read Section 4, we can take the definition of a class a little further and say that the characters of representation matrices from the same basis are the same for symmetry operations in the same class.) We have also seen that we can represent these symmetry operations in a mathematical manner. These mathematical representations are one of two types: *reducible* and *irreducible*. The reducible representations can be seen as combinations of irreducible representations, whereas the irreducible representations cannot be thought of as combinations of other irreducible representations. It turns out that the irreducible representations are the ones that we will be interested in as chemists. Accordingly, irreducible representations for all of the common point groups in chemistry have been calculated beforehand. These representations are given in tabular form, in a table known as a character table. The word *character* is used since the numbers are characteristic of the matrices from which they were derived. The character tables for the common chemical point groups are given in Appendix 1.

For example, the C_{2v} character table is shown below.

C_{2v}	E	C_2	$\sigma_v(xz)$	$\sigma_v'(yz)$		
A_1	1	1	1	1	z	x^2, y^2, z^2
A_2	1	1	-1	-1	R_z	xy
B_1	1	-1	1	-1	x, R_y	xz
B_2	1	-1	-1	1	y, R_x	yz

The layout of the character table is shown below. Notice that the irreducible representations are each given a symbol, known as a Mulliken symbol.

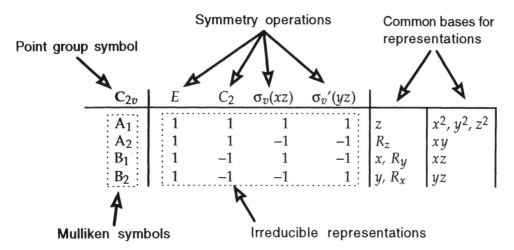

It is important that you become familiar with the layout and meaning of a character table, as we will use them again and again throughout the remainder of this text. For the moment the important thing to realise is that the irreducible representations have already been worked out for all of the point groups. We can, therefore, use the character tables much like any other piece of data, insofar as it can be looked up and used to solve problems in chemical group theory.

Let us break down the character table into its different parts.

1. The top left part of the table has the point group symbol.

2. On the same row and in the middle of the table are the symmetry operations of the group. They are grouped into classes (although the classes in C_{2v} all contain only one operation), with the number of symmetry elements in each class given before the symmetry element. For example, in C_{3v} (see Appendix 1) both $C_3{}^1$ and $C_3{}^2$ are in the same class, and appear in the character table as $2C_3$.

3. The point group's irreducible representations are then given below the symmetry operations. If you take a look at some of the character tables in Appendix 1, you will see that the irreducible representations do not necessarily only contain 1s and $-$1s. (N.B. for those who did not read Section 4, note that some of the representations do not appear to follow the rules that we discussed in Section 3 for generating a mathematical representation of a group; the reasons for this are explained in Section 4.) It is an important property of group theory that the total number of irreducible representations equals the number of classes of symmetry operations.

4. Each irreducible representation is given a shorthand symbol to describe it. These symbols are shown on the left-hand side of the table and are known as Mulliken symbols. For instance, in C_{2v} the 1 1 1 1 representation is given the Mulliken symbol A_1. There is a set of rules which can be used to assign Mulliken symbols to particular irreducible representations, but, for this text, it is sufficient to say that each irreducible representation

can be given a Mulliken symbol. The Mulliken symbol is simply a convenient shorthand method of writing the representation.

5.   The right-hand side of the table contains a variety of algebraic functions which occur commonly as bases for the generation of representations. We will use these later in the exercises.

5.3   The reduction formula

We have already seen (and described in Section 4) that representations known as reducible representations can be derived for particular bases of molecules. The actual practice of coming up with these reducible representations will be covered in the exercises in the following sections. It was also stated that each reducible representation can be broken down into a combination of irreducible representations. This part shows how this is done in practice. It may appear to be slightly mathematical, but the actual practice involves no more than very simple arithmetic.

The reduction formula—which comes straight from mathematical group theory, and will not be derived here—is shown below in its 'raw' form.

$$a_i = \left(\frac{1}{g}\right)\sum_R \left(\chi(R)^* \; \chi_i(R) \; n_R \right)$$

Where:

a_i = the number of times a particular irreducible representation appears in this reducible representation.
g = the number of symmetry operations in the point group.
$\chi(R)$ = the *character* of the reducible representation for a particular symmetry element.*
$\chi_i(R)$ = the *character* of the particular irreducible representation for a particular symmetry element.
n_R = the number of symmetry operations in that particular class.

All of this may appear to be a lot to take in, but it can be made clearer with an example.

The following is a reducible representation (RR) of the C_{3v} point group, along with the C_{3v} character table.

Reducible representation is:

	E	$2C_3$	$3\sigma_v$
RR	4	1	0

Character table for C_{3v} is:

C_{3v}	E	$2C_3$	$3\sigma_v$		
A_1	1	1	1	z	$x^2 + y^2, z^2$
A_2	1	1	−1	R_z	
E	2	−1	0	$(x,y)\;(R_x,R_y)$	$(x^2 - y^2, xy)$ (xz, yz)

The following method is used to figure out how many times the irreducible representation A_1 appears in the reducible representation.

---

* The star means that the complex conjugate of the irreducible representation should be used strictly here. However, for problems encountered in chemistry the representations containing complex numbers are rare, and this can be safely ignored.

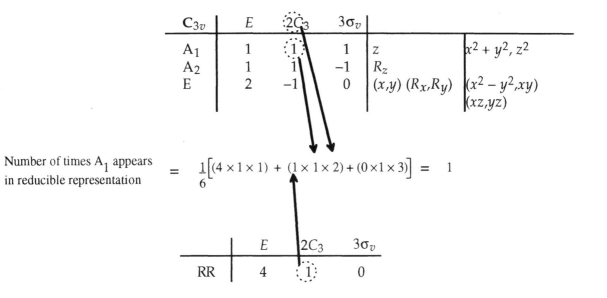

The equation is simple to use once you know where each part can be found. The $\frac{1}{6}$ part corresponds to the reciprocal of the total number of symmetry operations in the point group. This number is obtained by adding up all of the numbers of symmetry operations in each class. So, for C_{3v}, there is 1 operation in the E class, 2 operations in the C_3 class and 3 operations in the σ_v class: a total of 6. The next part of the equation works through the representation class by class. The picture above shows where each number comes from for the C_3 class and A_1 representation. The first number is the representation from the reducible representation for C_3, (1). The second number is the representation for that class in the A_1 irreducible representation (1), which can be read directly from the character table. The third number is the number of symmetry operations in that class (2). These three numbers are bracketed for convenience. The exercise is repeated for the E and σ_v classes. The total sum is equal to the number of times the A_1 representation appears in the reducible representation.

Using the reduction formula, figure out the number of times that the A_2 and E irreducible representations appear in the reducible representation.

Answer

Number of A_2 representations $= \dfrac{1}{6} [(4 \times 1 \times 1) + (1 \times 1 \times 2) + (0 \times -1 \times 3)] = 1$

Number of E representations $= \dfrac{1}{6} [(4 \times 2 \times 1) + (1 \times -1 \times 2) + (0 \times 0 \times 3)] = 1$

Thus, we have seen that using the reduction formula, we can reduce the reducible representation shown above to the following combination of irreducible representations:

$$A_1 + A_2 + E$$

The use of the reduction formula is quite general for reducing representations to their irreducible components. You must be familiar with the use of the formula, as it is used in nearly every group theory problem that you will encounter.

We will practise the use of the reduction formula again in the exercises which follow. Notice, however, that the formula itself is very easy to use, requiring no more than some simple arithmetic. It does require a little care in making sure that the correct numbers are transcribed to the correct part of the sum; this gets better with practice. Also notice that the answers to the formula must be integers; if they are not integers, then something has gone wrong with your arithmetic, or the reducible representation has not been constructed correctly.

Why do we go to such lengths to find out the irreducible representations of a particular representation? This question now leads us to the very heart of why group theory is of such importance in chemistry.

5.4   Schrödinger equation and group theory

We now reach the heart of group theory, and make the mathematical connection between the symmetry and energies of a molecule. In mathematical terms the 'energy states' of a molecule can be calculated using the Schrödinger equation, which in its simplest form is as follows:

$$H\psi = E\psi$$

where ψ is a mathematical function which describes the molecule, H is a mathematical *operator* which changes the ψ mathematical function into the mathematical function multiplied by its 'energy': $E\psi$.

There are certain rules which this equation follows, the most important of which is that only certain types of mathematical functions are solutions to this equation (so called eigenfunctions). In other words, we are restricted to only a handful of ψ functions which can be used to describe the molecule. The question is: what are these functions?

Let us ask ourselves a simple question. What is the effect on a molecule's energy if a symmetry operation (from the point group of the molecule) is carried out on it? Well, since the molecule after the symmetry operation is *indistinguishable* from the original, then the new molecule must have the same energy as the original. So, if we take a function ψ, which is a solution to the Schrödinger equation and perform a symmetry operation on it, then the new function, will also be a solution to the equation. This can be expressed more mathematically, if we say that the symmetry operation is denoted by the symbol O_S, then we can write the following:

$$HO_S\psi = EO_S\psi$$

This is a very important conclusion. What it says is that the functions (the ψs) which we chose to describe our molecule (whether they are based on wavefunctions, vectors or coordinates), must be a basis for the symmetry operations of the point group of the molecule. Otherwise, the above equation could not be true. This fact *greatly limits* the number of possible solutions to the Schrödinger equation, such that we can eliminate all but a few possibilities.

We have already seen the mathematical representations of O_S in the previous sections. It is these mathematical representations which tell us the possible solutions to the Schrödinger equation. In particular, it is the irreducible representations which have the most important meaning. The only solutions to a particular Schrödinger equation for a particular system must be a basis for an irreducible representation of the point group of the molecule. An example will, hopefully, make this clear.

Take the water molecule and consider the $2s$ and $2p$ atomic orbitals on the oxygen atom. How do these orbitals behave under the symmetry operations of the group?

What is the point group of the water molecule?

Answer

C_{2v}

Look up the symmetry operations of the C_{2v} point group. Sketch the atomic orbitals of the oxygen atom, and see how the orbitals change upon the symmetry operations of the group. Represent the result of the symmetry operation with +1 or −1. Then fill in the table below:-

	E	C_2	σ_v	σ_v
$2s$				
$2p_x$				
$2p_y$				
$2p_z$				

Answer

The orbitals transform as follows:

	E	C_2	σ_v	σ_v
$2s$	1	1	1	1
$2p_x$	1	−1	1	−1
$2p_y$	1	−1	−1	1
$2p_z$	1	1	1	1

All of these representations are the same as irreducible representations of the point group (see Appendix 1), and therefore we have shown that each of the atomic orbitals is a basis for the representation, and a possible solution to the Schrödinger equation which would describe these orbitals in water.

What about the hydrogen atoms? Pick a random basis and see if it forms a representation of the point group.

Consider the $1s$ orbital of just one of the hydrogen atoms. How does this orbital transform under the symmetry operations of the point group?

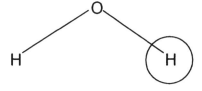

Using 1 or 0 to represent the orbital, fill in the following table. (Hint if the orbital is shifted away from its position assign a value of 0 to the representation.)

	E	C_2	σ_v	σ_v
$1s$				

Answer

	E	C_2	σ_v	σ_v'
$1s$	1	0	0	1

This representation clearly does not match any of the irreducible representations of the C_{2v} point group. Moreover, we cannot use the reduction formula to reduce this representation into anything smaller. Hence the *single* H $1s$ orbital in H_2O does *not* form a basis for a representation of the point group, and therefore cannot be used to calculate the energy of the molecule with the Schrödinger equation.

Where does this leave us? Well, consider the following possible basis for a representation of the group.

Now, consider both 1s orbitals of the hydrogen atoms, but as a linear combination. So, we now use $1s_a + 1s_b$ as a basis for our representation.

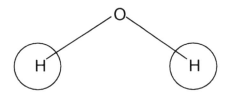

Fill in the following table, using +1 or −1 to represent how the combination of orbitals changes under the symmetry operations. (Hint: remember that $1s_a + 1s_b = 1s_b + 1s_a$)

	E	C_2	σ_v	σ_v
$1s_b + 1s_a$				

Answer

	E	C_2	σ_v	σ_v
$1s_b + 1s_a$	$1s_b + 1s_a$	$1s_a + 1s_b$	$1s_a + 1s_b$	$1s_b + 1s_a$
$1s_b + 1s_a$	1	1	1	1

So, the *combination* of the two orbitals *does form* an irreducible representation of the point group symmetry operations, and therefore, the *combination* of hydrogen $1s$ orbitals is a solution to the Schrödinger equation for these orbitals in H_2O. Notice that we must consider the two hydrogen atoms together, we cannot isolate them when it comes to talking about the molecule's energy. This, in a way, is common sense if we look at the molecule. It is clear that the two hydrogen atoms are symmetrically equivalent, and we would expect them to be 'linked' together in some way. Group theory allows us to make the link between our qualitative understanding of the symmetry of the molecule and a more quantitative description of the energy of the molecule.

Are there other combinations of the hydrogen $1s$ orbitals which can be used as a basis for an irreducible representation of the C_{2v} point group? [Hint: Ignore the trivial ones of $2(1s_b + 1s_a)$, they are simply the same as the one above.]

Answer

	E	C_2	σ_v	$\sigma_v{}'$
$1s_b - 1s_a$	$1s_b - 1s_a$	$1s_a - 1s_b$	$1s_a - 1s_b$	$1s_b + 1s_a$
$1s_b - 1s_a$	1	−1	−1	1

The 'out-of-phase' combination of $1s_b - 1s_a$ also forms a basis for an irreducible representation of the point group. Therefore, this combination is also a possible energy state for these orbitals in the water molecule.

Since $1s_a + 1s_b$ gives rise to the A_1 irreducible representation and $1s_a - 1s_b$ gives rise to the B_2 irreducible representation, then $1s_a + 1s_b$ is called the A_1 combination, and $1s_a - 1s_b$ is called the B_2 combination. We will use this shorthand form of describing orbitals from now on.

Moreover, the atomic orbitals which are on the oxygen atom give rise directly to an irreducible representation of the C_{2v} point group. Indeed, for atoms which lie on the 'point' of the group (i.e. are not changed in position by any of the symmetry operations) the atomic orbitals which give rise to particular irreducible representations are written in the columns at the right-hand side of the character table. Therefore, to determine the irreducible representations of atomic orbitals for an atom that lies on the point of the group, we simply need to look them up in the character table. For instance, in C_{2v} the p_x orbital of the oxygen atom (which is denoted by the letter x in the right-hand column of the character table) has B_1 symmetry. For the moment this is a useful point to remember and we will make use of it later.

In this short part, we have finally made the connection between the energy of a molecule and its symmetry. We have seen how the mathematical representations of the symmetry elements can be 'plugged' into the Schrödinger equation, with the immediate result that we greatly restrict the possible ways of describing the molecule. The ways of describing the molecule (using mathematical functions, such as most wavefunctions) *must* form a basis for the representation of the point group of the molecule. We can then describe these bases in terms of their symmetry. For instance, the p_x orbital on the O atom in H_2O has B_1 symmetry, and the $1s_a + 1s_b$ combination of hydrogen $1s$ orbitals has A_1 symmetry.

We will see in the following sections that the irreducible representations of our basis of a molecule are extremely useful in understanding the physical and chemical properties of the molecule.

5.5   Summary

In this section we have covered the following:

* The rules of a mathematical group can be applied to representations of molecular symmetry.

* A representation of molecular symmetry contains numbers which are called characters.

* The characters for reducible representations can be determined by following a few simple rules (see Section 6).

* Using the reduction formula allows us to 'reduce' the reducible representation into its irreducible components.

* The irreducible representations can be used to describe the functions which are solutions to the Schrödinger equation, since the possible functions must form a basis for the point group of the molecule.

* Using group theory we can determine how molecular symmetry greatly restricts the possible solutions to the Schrödinger equation.

SECTION 6

Group theory in action: molecular vibrations

Group theory in action: molecular vibrations

6.1   Some general points

We now turn our attention to the use of group theory in chemistry to solve problems related to the energy of the molecule. Our aim in all of these exercises is to come up with the irreducible representations which describe our molecular basis. We know from the previous section that these irreducible representations are real energy states of the molecule and, as such, they can tell us an enormous amount about the properties of the molecule.

The saying 'practice makes perfect' is true in doing group theory problems. The actual nitty-gritty of performing the exercises is almost trivial, but it is important that the exercises are performed accurately and completely; practice is the best method of becoming consistent and confident in completing the exercises.

In all of the exercises we will follow the same pattern, which is:

- decide on a basis to describe our molecule
- assign the point group of the molecule in question
- generate a reducible representation of our basis
- generate irreducible representations from the reducible representation
- examine the irreducible representations in terms of the molecular properties

Some new material will be introduced when it comes to examining the irreducible representations, but we will discuss that when it arises.

6.2   Molecular vibrations

6.2.1   A basis to describe vibrations

What is a molecular vibration? We can view all molecular vibrations as the relative motion of atoms with respect to one another. These vibrations, of course, are an energy state of the molecule, and as such we should be able to use group theory to figure out which vibrations the molecule can actually have.

To describe the relative motions of atoms, we need to set up a basis which can describe the motion of each individual atom. We have already seen this basis in Section 4, where the motion of each atom is represented by a set of orthogonal (i.e. at 90° to each other) displacement coordinates, as shown on the water molecule below.

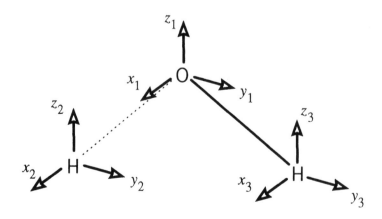

By sketching the ammonia molecule, draw a basis on the molecule that can be used to describe molecular vibrations.

Answer

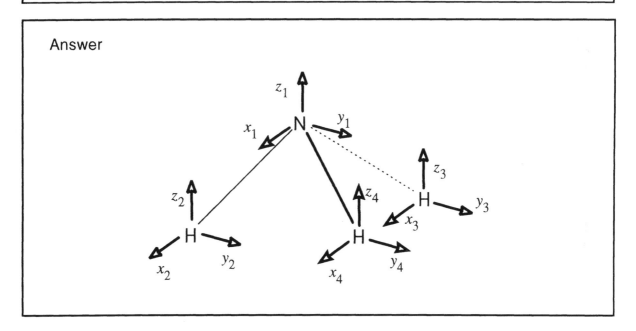

In the sketch above each atom has its own set of orthogonal cartesian displacement coordinates which can be used to describe the motion of that atom.

We will continue in this section with the water molecule, and there will be chance to examine other molecules in the exercises. The next step is to determine the point group of the water molecule. By now, you probably know the answer without needing the table at the end of Section 3; it is C_{2v}.

6.2.2 Generating a reducible representation

We now must generate a reducible representation of our basis. This we will reduce later, using the reduction formula, to obtain the relevant reducible representations. If you have not managed to read Section 4, then you must simply accept the rules to generate a reducible representation. For those that have read Section 4, we need to calculate the character of the matrix which represents the effect of each symmetry operation on the basis. Since we need only to figure out the sum of diagonal elements in the matrix to calculate its character, we need only to figure out how many atoms are unshifted by the symmetry operation, and then decide on the contribution of each coordinate to the character.

Two steps are required to generate a reducible representation for the sets of orthogonal cartesian coordinates shown above.

Step 1. For each symmetry operation in the group, figure out the number of *unshifted atoms* on for each symmetry operation.

Fill in the following table for the water molecule

C_{2v}	E	C_2	$\sigma_v(xz)$	$\sigma_v'(yz)$
No. of unshifted atoms				

Answer

C_{2v}	E	C_2	σ_v	σ_v'
No. of unshifted atoms	3	1	1	3

Step 2. Use the following equations to calculate the *contribution to the character* per unshifted atom of the representation.

$E = 3$ $i = -3$

$\sigma = 1$ $C_n = 1 + 2 \cos (360/n)°$ $S_n = -1 + 2 \cos (360/n)°$

Remember for C_{2v} $n = 2$.

Fill in the following table for the water molecule. (Hint: work through each symmetry operation, figuring out the contribution to the character. Some are trivial, like E, which is simply 3.)

C_{2v}	E	C_2	$\sigma_v(xz)$	$\sigma_v'(yz)$
No. of unshifted atoms	3	1	1	3
Contribution to character				

Answer

C_{2v}	E	C_2	σ_v	σ_v'
No. of unshifted atoms	3	1	1	3
Contribution to character	3	−1	1	1

To generate the reducible representation for the displacement coordinates, for each symmetry operation simply multiply the two numbers together. Notice that the reducible representation is given the symbol Γ_{3N} (Γ is an upper case Greek 'gamma'). The symbol is pronounced: 'gamma-three-en'.

Therefore, the reducible representation, Γ_{3N}, is given as follows:

C_{2v}	E	C_2	σ_v	σ_v'
No. of unshifted atoms	3	1	1	3
Contribution to character	3	−1	1	1
Γ_{3N}	9	−1	1	3

And, that is it! Our reducible representation is generated. Our next task is to find out how to reduce this representation into its irreducible parts.

6.2.3 Reducing the representation

Again, as above, we must follow a recipe to get to the answer. We need to use the reduction formula, which was explained above. When using the reduction formula, it is best to set-out your working in a simple and logical manner. If you do this then there is less chance of making silly mistakes in the addition. Use the method shown in Section 5 to help minimise the risk of errors.

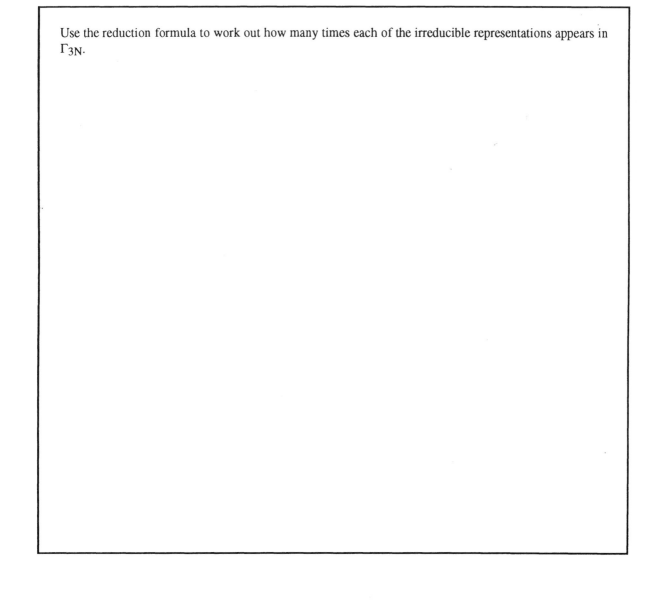

Use the reduction formula to work out how many times each of the irreducible representations appears in Γ_{3N}.

Answer

$$A_1 = \frac{1}{4}\left((9 \times 1 \times 1) + (-1 \times 1 \times 1) + (1 \times 1 \times 1) + (3 \times 1 \times 1)\right) = 3$$

$$A_2 = \frac{1}{4}\left((9 \times 1 \times 1) + (-1 \times 1 \times 1) + (1 \times -1 \times 1) + (3 \times -1 \times 1)\right) = 1$$

$$B_1 = \frac{1}{4}\left((9 \times 1 \times 1) + (-1 \times -1 \times 1) + (1 \times 1 \times 1) + (3 \times -1 \times 1)\right) = 2$$

$$B_2 = \frac{1}{4}\left((9 \times 1 \times 1) + (-1 \times -1 \times 1) + (1 \times -1 \times 1) + (3 \times 1 \times 1)\right) = 3$$

Therefore, $\Gamma_{3N} = 3A_1 + A_2 + 2B_1 + 3B_2$

Let us stop for a second to review what we have just done. We have generated a reducible representation of our description of a molecule. This reducible representation has then been reduced into its individual irreducible representation components. It is these irreducible representation components which are of interest to use, because these describe the possible energy states of the molecule according to our basis. Our next job is to figure out what these mean.

6.2.4 Examining the irreducible representations

Now think what possible answers we could obtain from a common sense point of view. Remember that our displacement coordinates tell us how each atom moves relative to the others in the molecule. Perhaps one immediate answer is that all the atoms could move in the same direction at once; this, of course, is the same as the translation of the whole molecule along an axis. There must be three of these translations, one along each of the cartesian axes, and they must be allowable solutions to the Schrödinger equation, because we know that molecules can move!

Indeed, our representation contains three irreducible representations which correspond to the 'symmetry' of translating the molecule along an axis. These have been worked out beforehand and are identified in the character table as x, y and z (see figure below).

C_{2v}	E	C_2	$\sigma_v(xz)$	$\sigma_v'(yz)$		
A_1	1	1	1	1	z	x^2, y^2, z^2
A_2	1	1	-1	-1	R_z	xy
B_1	1	-1	1	-1	x, R_y	xz
B_2	1	-1	-1	1	y, R_x	yz

We can see immediately, that the three translations belong to the A_1, B_1 and B_2 irreducible representations.

Furthermore, we can also envisage that certain combinations of displacement coordinates will give rise to rotations of the molecule, which, again, must be allowable solutions to the Schrödinger equation. There are three such rotations, each around a cartesian axis. The irreducible representations that the rotations belong to are given by R_x, R_y and R_z in the character table. We can see for C_{2v}, that the rotations belong to the A_2, B_1 and B_2 irreducible representations. Therefore, in our Γ_{3N} representation, six[*] of the irreducible representations correspond to translations and rotations of the molecule, and we can eliminate them from our Γ_{3N} representation.

---

[*] It is a general rule of molecular vibrations, that every non-linear molecule has $3N - 6$ vibrations, where N is the number of atoms. It is $3N$ because there are $3N$ possible displacement coordinates, and it is -6, because six 'displacements' correspond to rotations and translations. (For linear molecules, the number of vibrations is $3N - 5$.)

What we are left with is a collection of irreducible representations denoted by Γ_{vib} which *must* describe the symmetry of the vibrations within the molecule.

Therefore,

$$\Gamma_{vib} = \Gamma_{3N} - \Gamma_{T+R} \quad \text{(where T and R are translations and rotations)}$$

Calculate Γ_{vib} for the H_2O molecule.

Answer

$$\Gamma_{vib} = (3A_1 + A_2 + 2B_1 + 3B_2) - (A_1 + A_2 + 2B_1 + 2B_2)$$

$$\Gamma_{vib} = 2A_1 + B_2$$

So, we end up with the vibrations of the molecule described by the combination of irreducible representations, $2A_1 + B_2$. Since each irreducible representation corresponds to a single vibration,[*] we expect the water molecule to have three distinct vibrations—all other fundamental vibrations that we could think of do not exist, because they are not solutions to the Schrödinger equation.

Before we go on to examine the vibrations in more detail, it is worth repeating the exercise with a different molecule.

Using the basis sketched on page 79, find the irreducible representations for the vibrations of the NH_3 molecule.
(Hint: first determine the point group of the molecule, determine a reducible representation, reduce it, and then eliminate the translation and rotation contributions.)

(a) Reducible representation

---

[*] When we come to study E and T irreducible representations, we will see that these are double and triply degenerate, respectively. In other words, for molecular vibrations, an E representation corresponds to two vibrations at the same energy and a T representation corresponds to three vibrations at the same energy.

(b) Reduction formula

(c) Determination of Γ_{vib}

Answer

The point group of the molecule is C_{3v}.

(a) Reducible representation

	E	$2C_3$	$3\sigma_v$	
No. of unshifted atoms	4	1	2	
Contribution to character	3	0	1	(using formulae)
Γ_{3N}	12	0	2	

(b) Reduction formula (using character table)

$$A_1 = \frac{1}{6} \left((12 \times 1 \times 1) + (0 \times 1 \times 2) + (2 \times 1 \times 3) \right) = 3$$

$$A_2 = \frac{1}{6} \left((12 \times 1 \times 1) + (0 \times 1 \times 2) + (2 \times -1 \times 3) \right) = 1$$

$$E = \frac{1}{6} \left((12 \times 2 \times 1) + (0 \times -1 \times 2) + (2 \times 0 \times 3) \right) = 4$$

$$\Gamma_{3N} = 3A_1 + A_2 + 4E$$

(c) Determination of Γ_{vib}

From character table: $\Gamma_{T+R} = A_1 + A_2 + 2E$

[Notice that it is not 4E, because each E representation corresponds to a doubly degenerate state (i.e. two energy levels), therefore, 2E represents four energy states]

$$\Gamma_{vib} = \Gamma_{3N} - \Gamma_{T+R}$$

$$\Gamma_{vib} = (3A_1 + A_2 + 4E) - (A_1 + A_2 + 2E)$$

$$\Gamma_{vib} = 2A_1 + 2E$$

This result fits in with our $3N - 6$ rule, since there are four atoms, and we are expecting six vibrations. As the E representation is doubly degenerate, then 2E corresponds to four vibrations.

Now that we have determined both the symmetries and number of vibrations in a molecule, how can we use these to help explain the properties of the molecule? The first place to start is to ask, what do the vibrations actually look like? Can group theory help us to examine the vibrations in more detail? The answer is yes; with group theory we can determine what atomic motions the vibrations are actually made up of. To do this we need to use a technique known as internal coordinates and projection operators.

6.3 Internal coordinates

Turning back to the water molecule, we determined that the internal vibrations of the molecule correspond to the irreducible representations of $2A_1$ and B_2. But, we are faced with a problem: of these vibrations, which are bond-stretching vibrations and which are bending vibrations? There is no immediate way of telling. To overcome this, we need to look back at our basis of $3N$ displacement coordinates. This basis gives us all the vibrations of the molecule, including the translations and rotations, but this is almost too much information. What happens when we restrict the basis to something simpler?

Consider a basis where all we are interested in is the stretching vibrations of the molecule. A stretching vibration can be considered as the displacement of one atom with respect to another along the line of the bond that joins them. So, for instance, we can define two new displacement coordinates for the water molecule (r_1 and r_2), which lie along the O–H bonds.

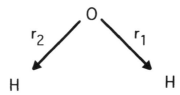

These displacement coordinates, which are called internal coordinates, can be used as a basis for a reducible representation. In fact, coming up with a reducible representation for this basis is made easier, since the contribution to the character in all cases is 1. (Notice that this differs from when we used the 3N basis.) Therefore, to figure out a reducible representation for this basis, all we have to do is figure out the number of unshifted displacement coordinates (not atoms!) upon each symmetry operation. We also do not need to remove contributions from translations and rotations either, since our basis cannot accommodate these possibilities.

Complete the following exercise:

Using internal coordinates as a basis, derive a reducible representation for the stretching vibrations in the H_2O molecule. Reduce the representation. (Use the nomenclature 'Γ_{str}' for the representation.)

Answer

Point group is C_{2v}.

C_{2v}	E	C_2	σ_v	σ_v'
No. of unshifted coordinates	2	0	0	2
Γ_{str}	2	0	0	2

$$A_1 = \frac{1}{4}((2 \times 1 \times 1) + (0 \times 1 \times 1) + (0 \times 1 \times 1) + (2 \times 1 \times 1)) = 1$$

$$A_2 = \frac{1}{4}((2 \times 1 \times 1) + (0 \times 1 \times 1) + (0 \times -1 \times 1) + (2 \times -1 \times 1)) = 0$$

$$B_1 = \frac{1}{4}((2 \times 1 \times 1) + (0 \times -1 \times 1) + (0 \times 1 \times 1) + (2 \times -1 \times 1)) = 0$$

$$B_2 = \frac{1}{4}((2 \times 1 \times 1) + (0 \times -1 \times 1) + (0 \times -1 \times 1) + (2 \times 1 \times 1)) = 1$$

Therefore, $\Gamma_{str} = A_1 + B_2$

We can immediately see, that we expect two stretching vibrations for H_2O; one with A_1 symmetry and one with B_2 symmetry. If we also compare Γ_{str} with Γ_{vib}, we see that $\Gamma_{vib} = 2A_1 + B_2$. Since, $A_1 + B_2$ corresponds to stretching vibrations, the remainder, A_1, must correspond to a single bending vibration. (This fits in, since from a qualitative point of view, we can imagine only one possible bending vibration in the H_2O molecule.) So, we can write the following:

$$\Gamma_{str} = A_1 + B_2$$
$$\Gamma_{bend} = A_1$$

For the NH_3 molecule, use internal coordinates to determine the stretching vibrations in the molecule. Using the result for Γ_{vib} on page 85, determine Γ_{bend}.

Answer

The point group of the molecule is C_{3v}.

a) Reducible representation

	E	$2C_3$	$3\sigma_v$
No. of unshifted coordinates (or bonds)	3	0	1
Γ_{str}	3	0	1

b) Reduction formula (using character table)

$$A_1 = \frac{1}{6}\left((3 \times 1 \times 1) + (0 \times 1 \times 2) + (1 \times 1 \times 3)\right) = 1$$

$$A_2 = \frac{1}{6}\left((3 \times 1 \times 1) + (0 \times 1 \times 2) + (1 \times -1 \times 3)\right) = 0$$

$$E = \frac{1}{6}\left((3 \times 2 \times 1) + (0 \times -1 \times 2) + (1 \times 0 \times 3)\right) = 1$$

$$\Gamma_{str} = A_1 + E$$

c) Determination of Γ_{bend}

From above: $\Gamma_{vib} = 2A_1 + 2E$

$\Gamma_{bend} = \Gamma_{vib} - \Gamma_{str}$

$\Gamma_{bend} = A_1 + E$

Therefore, we can determine Γ_{bend} from the difference of Γ_{vib} and Γ_{str}. Can we determine Γ_{bend} by using internal coordinates, in a similar way to the determination of Γ_{str}? Try the following.

For the NH_3 molecule, can you think of some internal coordinates which could be used as a basis for determining Γ_{bend}? Sketch the basis on a drawing of the NH_3 molecule.

Answer

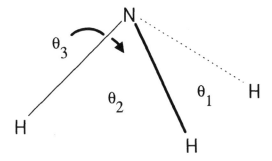

The basis consists of the angle formed between the bonds.

Using the angles as an internal coordinate basis, derive a reducible representation for the basis. (Hint, if an angle is 'reversed' by a symmetry operation, then it is counted as unshifted. Also, these internal coordinates behave in exactly the same way as displacement coordinates, insofar as all we have to do is figure out how many are unshifted by the symmetry operation.) Reduce the representation.

Answer

The point group of the molecule is C_{3v}.

a) Reducible representation

	E	$2C_3$	$3\sigma_v$
No. of unshifted angles	3	0	1
Γ_{bend}	3	0	1

b) Reduction formula (using character table)

$$A_1 = \frac{1}{6}\left((3 \times 1 \times 1) + (0 \times 1 \times 2) + (1 \times 1 \times 3)\right) = 1$$

$$A_2 = \frac{1}{6}\left((3 \times 1 \times 1) + (0 \times 1 \times 2) + (1 \times -1 \times 3)\right) = 0$$

$$E = \frac{1}{6}\left((3 \times 2 \times 1) + (0 \times -1 \times 2) + (1 \times 0 \times 3)\right) = 1$$

$$\Gamma_{bend} = A_1 + E$$

Exactly the same answer as we got from the difference method. (CAUTION: Sometimes, generating Γ_{bend} from the angles of the molecule, does NOT give the same answer as the difference method. Now and then, a *redundant coordinate* is found in the Γ_{bend} representation, which is often A_1. When this occurs the redundant coordinate can be simply removed from the Γ_{bend} representation. One such case occurs in the problem at the end of this section.)

Let us recap on where we have got to. We can now determine the symmetries of the possible bending and stretching vibrations for any molecule. To do this we select a basis with which to describe the molecular vibrations. This basis can try to encompass all molecular motions ($3N$ basis) or specific types of vibration, depending on the properties of the molecule we wish to examine.

Now that we have the symmetries of the particular vibrations, can we use group theory to analyse the vibrations further, and also provide a link with experiment?

6.4   Projection operators

This is perhaps the most satisfying part of group theory. We have done the hard work in determining the symmetry of particular molecular vibrations. We can now use these results and use group theory to let us 'see' what the vibrations look like. In other words, for a particular vibration we can determine how each atom is moving. The mathematical technique that allows us to do this is called a projection operator.

What is the significance of this? What we have determined so far is the symmetry of particular vibrations. For instance, we have determined that the water molecule has two possible stretching vibrations, one with symmetry A_1 and the other with symmetry B_2. A vibration with 'symmetry A_1' means that the linear combination of the coordinates which go to make up its basis also have A_1 symmetry (a linear combination is simply the addition/subtraction of the coordinates, e.g. $r_1 + r_2$). We can use the projection operator to tell us what the linear combination is, which in turn represents the atomic motions.

The mathematical roots of the projection operator are beyond this text. However, as with the previous parts of group theory, the use of the projection operator is almost trivial, despite the rather daunting equation which is given at the top of the next page.

For a particular irreducible representation:

$$Px = \left(\sum_R [\chi(R)\,R]\right)x$$

P is the projection operator
x is the generating function, coordinate or vector
$\chi(R)$ is the character of the irreducible representation for the symmetry operation R
R is the symmetry operation

An example will, hopefully, make all of this clear.

Go back to the water molecule. Consider only the stretching vibrations, and thus, consider the internal coordinates, r_1 and r_2, which describe the bond stretches. We need to choose one of these to be our generating coordinate, say r_1. (This is x in the equation above.)

The equation tells us that for each symmetry operation, we need to say what r_1 is converted into. We do this simply by drawing up the following table:

C_{2v}	E	C_2	σ_v	$\sigma_v{'}$
r_1	r_1	r_2	r_2	r_1

The next part of the equation tells us that we must now take each of the irreducible representations which the stretching vibrations correspond to (A_1 and B_2), and multiply the representations with the results of performing symmetry operations on the generating coordinate.

So, for the A_1 representation, which is as follows:-

C_{2v}	E	C_2	σ_v	σ_v
A_1	1	1	1	1

Combining the two tables, gives:

C_{2v}	E	C_2	σ_v	$\sigma_v{'}$
	$1 \times r_1$	$1 \times r_2$	$1 \times r_2$	$1 \times r_1$

Adding each part together gives:

$r_1 \rightarrow r_1 + r_2 + r_1 + r_2 \quad = \quad 2r_1 + 2r_2$

Since we are only interested in the *relative* way in which vectors combine,[*] then we can say that the A_1 representation, has $r_1 + r_2$ as a basis. What does this mean? It tells us that the A_1 stretch has r_1 and r_2 increasing or decreasing at the same time. Or, in other words, as the r_1 coordinate increases so does the r_2 coordinate, and as the r_1 coordinate decreases so does the r_2 coordinate. The figure below shows this:

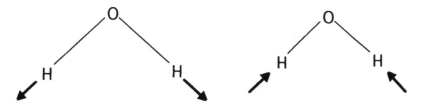

---

[*] Strictly we should normalise the linear combinations.

$r_1 + r_2$ is known as a *symmetry adapted linear combination* of the displacement coordinates. From now on this term will be abbreviated to SALC. Note here that we have not normalised the SALC, which in its full form is given as: $\frac{1}{\sqrt{2}}(r_1 + r_2)$.

Using the projection operator method, show the motion of the atoms in the B_2 vibration in water, sketch the result.

Answer

Use r_1 as a generating coordinate.

C_{2v}	E	C_2	σ_v	σ_v'
r_1	r_1	r_2	r_2	r_1

Multiply by B_2 representation

C_{2v}	E	C_2	σ_v	σ_v'
B_2	1	-1	-1	1
	r_1	$-r_2$	$-r_2$	r_1

$r_1 \rightarrow r_1 - r_2 - r_2 + r_1 = 2r_1 - 2r_2$ which is equivalent to $r_1 - r_2$.

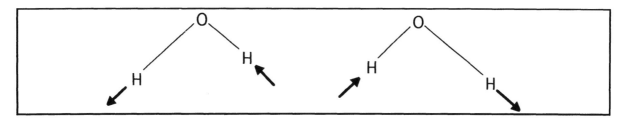

The result shows that the B_2 vibration in water is the one where the $\mathbf{r}_1$ coordinate is reducing whilst the $\mathbf{r}_2$ coordinate is increasing, and vice versa. (To maintain the centre of mass in the molecule, there is also a small net motion of the oxygen atom along the y axis in this vibration.)

We now have a complete vibrational picture of the H_2O molecule. Unsurprisingly, the A_1 vibration is known as the symmetric stretch, and the B_2 vibration is known as the antisymmetric stretch. The remaining A_1 bend, is simply the bending motion of the molecule.

Let us turn our attention to the NH_3 molecule.

Using the projection operator method determine the SALCs for the stretching vibrations in NH_3 (point group C_{3v}). Remember that we have already determined $\Gamma_{str} = A_1 + E$. Sketch the result. It was not apparent in the example above, but we must use the 'full' form of the character table for the projection operator exercise; this is shown below. (N.B. Note that there is an apparent problem with this analysis, insofar as the projection operator only gives a single result for the E vibration, which should have two vibrational motions. For the moment ignore this, as we will discuss it immediately below.)

Use the following 'full' symmetry operations: $E \quad C_3{}^1 \quad C_3{}^2 \quad \sigma_v(1) \quad \sigma_v(2) \quad \sigma_v(3)$.

Answer

The set of internal coordinates as a basis for the stretching vibrations is as follows:

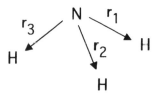

Using r_1 as a generating coordinate:

C_{3v}	E	$C_3{}^1$	$C_3{}^2$	$\sigma_v(1)$	$\sigma_v(2)$	$\sigma_v(3)$
r_1	r_1	r_2	r_3	r_1	r_3	r_2

Multiplying by the appropriate characters for A_1 and E, gives:

C_{3v}	E	$C_3{}^1$	$C_3{}^2$	$\sigma_v(1)$	$\sigma_v(2)$	$\sigma_v(3)$
A_1	$1 \times r_1$	$1 \times r_2$	$1 \times r_3$	$1 \times r_1$	$1 \times r_3$	$1 \times r_2$
E	$2 \times r_1$	$-1 \times r_2$	$-1 \times r_3$	$0 \times r_1$	$0 \times r_3$	$0 \times r_2$

A_1 is equivalent to $r_1 + r_2 + r_3$
E is equivalent to $2r_1 - r_2 - r_3$

(again, both forms are not normalised)

In sketch form, these appear as:

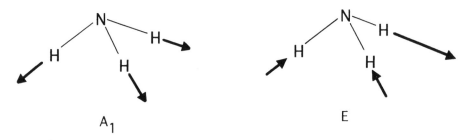

Notice for the E vibration one of the bonds is stretching twice as far as the other two bonds.

Now we turn to our problem. The E vibration, of course, is doubly degenerate and should have two different types of vibrational motions. So far, we have only discovered one. If we were to use another generating coordinate, say r_2, then we would end up with a vibration that looks essentially the same as the one generated from r_1. This result does not help us in deciding what the second E vibration 'looks like'. To solve the problem we need to use a different type of generating coordinate. Unfortunately, it is beyond this text to explain why we need to do this, but for those who are interested more advanced books on group theory have the answer, which is to do with the orthogonality of the generating coordinates. Choosing $r_2 - r_3$ as a generating coordinate, we get the following:

C_{3v}	E	$C_3{}^1$	$C_3{}^2$	$\sigma_v(1)$	$\sigma_v(2)$	$\sigma_v(3)$
$r_2 - r_3$	$2(r_2 - r_3)$	$r_2 - r_1$	$r_1 - r_3$	0	0	0

which gives $3(r_2 - r_3)$, which is equivalent to $r_2 - r_3$

Therefore, the second E vibration looks like

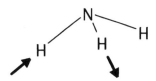

The problem of finding a suitable generating coordinate for the projection operator method happens in all point groups where there are doubly degenerate (E) and triply degenerate (T) representations.

For the NH_3 molecule use the projection operator method to determine the SALCs for the *bending* vibrations of the molecule. On page 88 we determined that $\Gamma_{bend} = A_1 + E$. Use $(\theta_2 - \theta_3)$ as a generating coordinate for the second E vibration.

Answer

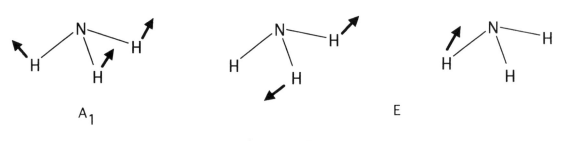

Using θ_1 as a generating coordinate, we obtain the following.

$\mathbf{C}_{3v}$	E	$C_3{}^1$	$C_3{}^2$	$\sigma_v(1)$	$\sigma_v(2)$	$\sigma_v(3)$
θ_1	θ_1	θ_2	θ_3	θ_3	θ_1	θ_2

Multiplying by the appropriate characters for A_1 and E, gives:-

$\mathbf{C}_{3v}$	E	$C_3{}^1$	$C_3{}^2$	$\sigma_v(1)$	$\sigma_v(2)$	$\sigma_v(3)$
A_1	$1 \times \theta_1$	$1 \times \theta_2$	$1 \times \theta_3$	$1 \times \theta_3$	$1 \times \theta_1$	$1 \times \theta_2$
E	$2 \times \theta_1$	$-1 \times \theta_2$	$-1 \times \theta_3$	$0 \times \theta_3$	$0 \times \theta_1$	$0 \times \theta_2$

A_1 is equivalent to $\qquad \theta_1 + \theta_2 + \theta_3$
E is equivalent to $\qquad 2\theta_1 - \theta_2 - \theta_3$

(again, both forms are not normalised).

Using $\theta_2 - \theta_3$ as a generating coordinate, we obtain the following.

$\mathbf{C}_{3v}$	E	$C_3{}^1$	$C_3{}^2$	$\sigma_v(1)$	$\sigma_v(2)$	$\sigma_v(3)$
$\theta_2 - \theta_3$	$\theta_2 - \theta_3$	$\theta_3 - \theta_1$	$\theta_1 - \theta_2$	$\theta_3 - \theta_2$	$\theta_2 - \theta_1$	$\theta_1 - \theta_3$

Multiplying by the appropriate characters for A_1 and E, gives:

$\mathbf{C}_{3v}$	E	$C_3{}^1$	$C_3{}^2$
E	$2 \times (\theta_2 - \theta_3)$	$-1 \times (\theta_3 - \theta_1)$	$-1 \times (\theta_1 - \theta_2)$

E is equivalent to $(3\theta_2 - 3\theta_3)$, which is equivalent to $(\theta_2 - \theta_3)$

Therefore, the three SALCs are:

A_1 is equivalent to $\qquad \theta_1 + \theta_2 + \theta_3$
E is equivalent to $\qquad 2\theta_1 - \theta_2 - \theta_3$ and $\theta_2 - \theta_3$

In sketch form, these appear as:

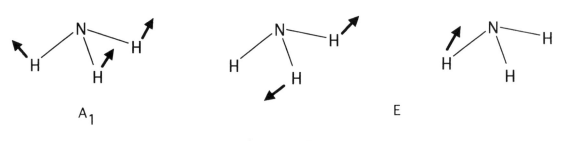

Assign a point group to $AuBr_4^-$

Answer

$\mathbf{D}_{4h}$

For $AuBr_4^-$, determine the SALCs for the stretching vibrations of the molecule. Use the projection operator method to 'view' the SALCs, sketch the results. (Hint: you need to determine Γ_{str} for this exercise, using internal coordinates as a basis. Use $\mathbf{r}_2$ as the generating coordinate for the second E vibration.)

Use the 'full' form of the character table for the projection operators, which is:-

$\mathbf{D}_{4h}$ E C_4^1 C_4^3 C_2 $C_2'(x)$ $C_2'(y)$ $C_2''(1)$ $C_2''(2)$ i S_4^1 S_4^3 σ_h $\sigma_v(x)$ $\sigma_v(y)$ $\sigma_d(1)$ $\sigma_d(2)$

$C_2(x)$ corresponds to the C_2 axis along one of the Au–Br bonds. $C_2''(1)$ corresponds to one of the C_2 axes that lies between the Au–Br bonds. C_2 is the axis which is co-axial with the principal axis.

Answer

Using internal coordinates as a basis for a reducible representation, we obtain.
(See character table in Appendix 1.)

$\mathbf{D}_{4h}$	E	$2C_4$	C_2	$2C_2'$	$2C_2''$	i	$2S_4$	σ_h	$2\sigma_v$	$2\sigma_d$
No. unshifted	4	0	0	2	0	0	0	4	2	0
Γ_{str}	4	0	0	2	0	0	0	4	2	0

Using the reduction formula

$$A_{1g} = \frac{1}{16}\left(\begin{array}{l} (4\times1\times1) + (0\times1\times2) + (0\times1\times1) + (2\times1\times2) + (0\times1\times2) \\ + (0\times1\times1) + (0\times1\times2) + (4\times1\times1) + (2\times1\times2) + (0\times1\times2) \end{array} \right) = 1$$

$$A_{2g} = \frac{1}{16}\left(\begin{array}{l} (4\times1\times1) + (0\times1\times2) + (0\times1\times1) + (2\times-1\times2) + (0\times-1\times2) \\ + (0\times1\times1) + (0\times1\times2) + (4\times1\times1) + (2\times-1\times2) + (0\times-1\times2) \end{array} \right) = 0$$

$$B_{1g} = \frac{1}{16}\left(\begin{array}{l} (4\times1\times1) + (0\times-1\times2) + (0\times1\times1) + (2\times1\times2) + (0\times-1\times2) \\ + (0\times1\times1) + (0\times-1\times2) + (4\times1\times1) + (2\times1\times2) + (0\times-1\times2) \end{array} \right) = 1$$

$$B_{2g} = \frac{1}{16}\left(\begin{array}{l} (4\times1\times1) + (0\times-1\times2) + (0\times1\times1) + (2\times-1\times2) + (0\times1\times2) \\ + (0\times1\times1) + (0\times-1\times2) + (4\times1\times1) + (2\times-1\times2) + (0\times1\times2) \end{array} \right) = 0$$

$$E_g = \frac{1}{16}\left(\begin{array}{l} (4\times2\times1) + (0\times0\times2) + (0\times-2\times1) + (2\times0\times2) + (0\times0\times2) \\ + (0\times2\times1) + (0\times0\times2) + (4\times-2\times1) + (2\times0\times2) + (0\times0\times2) \end{array} \right) = 0$$

$$A_{1u} = \frac{1}{16}\left(\begin{array}{l} (4\times1\times1) + (0\times1\times2) + (0\times1\times1) + (2\times1\times2) + (0\times1\times2) \\ + (0\times-1\times1) + (0\times-1\times2) + (4\times-1\times1) + (2\times-1\times2) + (0\times-1\times2) \end{array} \right) = 0$$

$$A_{2u} = \frac{1}{16}\left(\begin{array}{l} (4\times1\times1) + (0\times1\times2) + (0\times1\times1) + (2\times-1\times2) + (0\times-1\times2) \\ + (0\times-1\times1) + (0\times-1\times2) + (4\times-1\times1) + (2\times1\times2) + (0\times1\times2) \end{array} \right) = 0$$

$$B_{1u} = \frac{1}{16}\left(\begin{array}{l} (4\times1\times1) + (0\times-1\times2) + (0\times1\times1) + (2\times1\times2) + (0\times-1\times2) \\ + (0\times-1\times1) + (0\times1\times2) + (4\times-1\times1) + (2\times-1\times2) + (0\times1\times2) \end{array} \right) = 0$$

$$B_{2u} = \frac{1}{16}\left(\begin{array}{l} (4\times1\times1) + (0\times-1\times2) + (0\times1\times1) + (2\times-1\times2) + (0\times1\times2) \\ + (0\times-1\times1) + (0\times1\times2) + (4\times-1\times1) + (2\times1\times2) + (0\times-1\times2) \end{array} \right) = 0$$

$$E_u = \frac{1}{16}\left(\begin{array}{l} (4\times2\times1) + (0\times0\times2) + (0\times-2\times1) + (2\times0\times2) + (0\times0\times2) \\ + (0\times-2\times1) + (0\times0\times2) + (4\times2\times1) + (2\times0\times2) + (0\times0\times2) \end{array} \right) = 1$$

Which gives $\Gamma_{str} = A_{1g} + B_{1g} + E_u$

To determine the SALCs we need to use a generating coordinate. In the diagram below r_1 can be used as a generating coordinate.

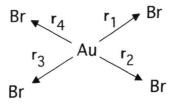

If we do this:

D_{4h}	E	C_4^1	C_4^3	C_2	$C_2'(x)$	$C_2'(y)$	$C_2''(1)$	$C_2''(2)$	i	S_4^1	S_4^3	σ_h	$\sigma_v(x)$	$\sigma_v(y)$	$\sigma_d(1)$	$\sigma_d(2)$
r_1	r_1	r_2	r_4	r_3	r_1	r_3	r_2	r_4	r_3	r_2	r_4	r_1	r_1	r_3	r_4	r_2
A_{1g}	r_1	r_2	r_4	r_3	r_1	r_3	r_2	r_4	r_3	r_2	r_4	r_1	r_1	r_3	r_4	r_2
B_{1g}	r_1	$-r_2$	$-r_4$	r_3	r_1	r_3	$-r_2$	$-r_4$	r_3	$-r_2$	$-r_4$	r_1	r_1	r_3	$-r_4$	$-r_2$
E_u	$2r_1$	$0r_2$	$0r_4$	$-2r_3$	$0r_1$	$0r_3$	$0r_2$	$0r_4$	$-2r_3$	$0r_2$	$0r_4$	$2r_1$	$0r_1$	$0r_3$	$0r_4$	$0r_2$

which gives

A_{1g} is equivalent to: $4r_1 + 4r_2 + 4r_3 + 4r_4$, which is equivalent to $r_1 + r_2 + r_3 + r_4$
B_{1g} is equivalent to: $4r_1 - 4r_2 + 4r_3 - 4r_4$, which is equivalent to $r_1 - r_2 + r_3 - r_4$
E_u is equivalent to: $4r_1 - 4r_3$, which is equivalent to $r_1 - r_3$

For the remaining E_u SALC, use r_2 as a generating coordinate

D_{4h}	E	C_4^1	C_4^3	C_2	$C_2'(x)$	$C_2'(y)$	$C_2''(1)$	$C_2''(2)$	i	S_4^1	S_4^3	σ_h	$\sigma_v(x)$	$\sigma_v(y)$	$\sigma_d(1)$	$\sigma_d(2)$
r_2	r_2	r_3	r_1	r_4	r_4	r_2	r_1	r_3	r_4	r_3	r_1	r_2	r_4	r_2	r_3	r_1
E_u	$2r_2$	$0r_3$	$0r_1$	$-2r_4$	$0r_4$	$0r_2$	$0r_1$	$0r_3$	$-2r_4$	$0r_3$	$0r_1$	$2r_2$	$0r_4$	$0r_2$	$0r_3$	$0r_1$

which gives

E_u is equivalent to: $4r_2 - 4r_4$, which is equivalent to $r_2 - r_4$

In picture form:

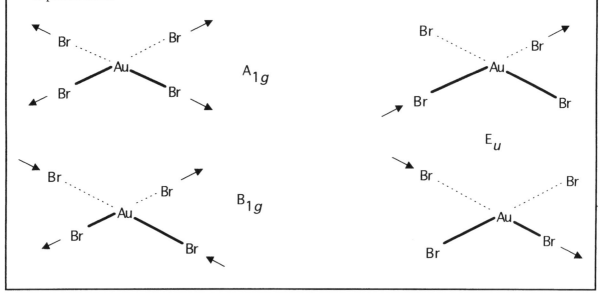

Hopefully the exercises have shown that projection operators are very useful for seeing how group theory is important in studying the vibrations of molecules. Can we now take this one stage further, and see if the group theory predictions are borne out in experiment? The experimental techniques to study molecular vibrations are infra-red and Raman spectroscopy. Does an infra-red spectrum of a simple molecule match what we predict from group theory? The next part attempts to answer this question.

6.5   Spectroscopy and symmetry selection rules

As with any form of spectroscopy, to observe a band in a spectrum a transition must occur between energy states. So far, group theory has told us what vibrational energy states exist and also tells us the 'symmetries' of those states, but it tells us nothing about whether it is possible to observe a transition to any of these states from the ground state of the molecule. Whether a transition is allowed or not can be determined by solving the following equation.

$$\int \psi_i^* \, \text{T} \, \psi_f \, \text{d}\tau$$

Where: ψ_i is the wavefunction of the ground state;[†]
T is the transition moment operator;
ψ_f is the wavefunction of the excited state;
$\text{d}\tau$ means that the integration is carried out over all space.

For a transition to be allowed the equation above must not equal zero. This looks like a daunting equation to have to solve (or even look at!). Fortunately, we can use the results of group theory to simplify the use of this equation greatly, with no need to know any serious mathematics. All that is required is that we follow certain rules to arrive at the answer.

6.5.1   Infra-red spectroscopy

There are two forms of commonly used vibrational spectroscopy; these are infra-red spectroscopy and Raman spectroscopy. In infra-red spectroscopy, a molecule can be excited from its ground vibrational state to a fundamental vibration using infra-red radiation. This all occurs if there is a dipole moment change in the molecule upon absorption. So, for infra-red spectroscopy, the equation above can be rewritten as:

$$\int \psi_i \, \mu \, \psi_f \, \text{d}\tau$$

Where: ψ_i is the wavefunction of the ground vibrational state;
μ is the dipole moment of the molecule;
ψ_f is the wavefunction of the vibrational excited state.

How can group theory help us here? Take the water molecule as an example above. We have already calculated that the water molecule has three excited fundamental vibrational states, which have the symmetries: A_1, A_1 and B_2. Therefore, for each of the three excited states, we could represent the ψ_f wavefunction with the its symmetry. For example, if we were interested in the A_1 vibrations we could replace ψ_f with A_1 in the above equation, thus:

$$\int \psi_i \, \mu \, A_1 \, \text{d}\tau$$

What about ψ_i and μ? For any ground vibrational state the symmetry of that state is always A_1, for any molecule. So, we can replace ψ_i with A_1. In the case of the dipole moment, μ, of the molecule, it turns out that it always has the same symmetry as the x, y and z translations of a molecule. To find out the symmetry of μ, we need simply to turn to the appropriate character table for the molecule, and see which irreducible representations x, y and z belong to.

Turn to the C_{2v} character table and write down the irreducible representations which x, y and z belong to.

[†] Strictly we must use the complex conjugate of the wavefunction for this equation to make sense, hence the use of ψ_i^* instead of just ψ_i.

Answer

x — B_1, y — B_2 and z — A_1

So we can now rewrite the equation in terms of its symmetry, as follows:

$$\int A_1 \begin{pmatrix} B_1 \\ B_2 \\ A_1 \end{pmatrix} A_1 \; d\tau$$

This equation is made even simpler, since all we have to decide is whether the symmetries of the appropriate parts of the equation all multiply together to give a non-zero result, without having to do the integration: in group theory language all we need to decide is whether the multiplication of the symmetries gives the totally symmetric representation (e.g. A_1, A_{1g}) or not. So, all we are left to figure out is the result of the following:

$$A_1 \times B_1 \times A_1$$
$$A_1 \times B_2 \times A_1$$
$$A_1 \times A_1 \times A_1$$

To multiply irreducible representations together, we need to use the rules given in Appendix 2: using these rules:

$$A_1 \times B_1 \times A_1 = B_1$$
$$A_1 \times B_2 \times A_1 = B_2$$
$$A_1 \times A_1 \times A_1 = A_1$$

We see immediately that A_1 is included in one of the answers, and we can conclude that absorbances due to transitions from the ground state to the A_1 vibrations in the water molecule will be observed in the infra-red spectrum. (There are two A_1 vibrations, a bend and a stretch.)

Will the B_2 vibration in water be observable by infra-red spectroscopy?

The symmetry of the ground state is A_1. The symmetries of the dipole moment of the molecule correspond to the symmetries of the x, y and z irreducible representations in C_{2v}, which can be read directly from the character table and are B_1, B_2 and A_1. The excited state has B_2 symmetry. Hence we can write the following:

$$A_1 \times B_1 \times B_2 = A_2$$
$$A_1 \times B_2 \times B_2 = A_1$$
$$A_1 \times A_1 \times B_2 = B_2$$

One of the answers contains the A_1 representation, hence we expect to observe an absorption in the infra-red spectrum of H_2O that corresponds to a transition from the ground vibrational state to the B_2 vibrational state.

6.5.2 Raman spectroscopy

How about Raman spectroscopy? To determine whether or not we expect to observe an absorption in the Raman spectrum due to a vibrational transition, we can use the same technique as we did for infra-red spectroscopy above. The difference is that instead of using the symmetries of the dipole moment, we need to figure out the symmetries of the polarisability tensor* of the molecule. It turns out that the tensor has the same symmetry as x^2, y^2, z^2, xy, zx and yz, or any combination of these. Again, we can look up these symmetries in the appropriate character table.

Going back to the water molecule. Are transitions to the A_1 vibrations observed in the Raman spectrum? Turning to the C_{2v} character table, we see that x^2, y^2 and z^2 have A_1 symmetry, xz has B_1 symmetry, yz has B_2 symmetry, and xy has A_2 symmetry. Hence, we can write the following:-

$$A_1 \times \begin{pmatrix} B_1 \\ B_2 \\ A_1 \\ A_2 \end{pmatrix} \times A_1$$

Which can be rewritten as:

$$A_1 \times B_1 \times A_1 = B_1$$
$$A_1 \times B_2 \times A_1 = B_2$$
$$A_1 \times A_1 \times A_1 = A_1$$
$$A_1 \times A_2 \times A_1 = A_2$$

One equation contains the A_1 representation, hence we expect to see two bands in the Raman spectrum corresponding to the A_1 vibrations (one for each A_1 vibration, remember that there are two A_1 vibrations in water: the symmetric stretch and the bending mode).

Will the B_2 vibration in water be observable by Raman spectroscopy?

The symmetry of the ground state is A_1. The excited state has B_2 symmetry. Hence we can write the following:

$$A_1 \times B_1 \times B_2 = A_2$$
$$A_1 \times B_2 \times B_2 = A_1$$
$$A_1 \times A_1 \times B_2 = B_2$$
$$A_1 \times A_2 \times B_2 = B_1$$

One of the answers contains the A_1 representation, hence we expect to observe an absorption in the Raman spectrum of H_2O, which corresponds to a transition from the ground vibrational state to the B_2 vibrational state.

If a band due to a vibration appears in the infra-red spectrum, the vibration is known as *infra-red active* and if a band appears in the Raman spectrum, the vibration is known as *Raman active*. Therefore, for the water molecule, we expect to see three fundamental vibration bands, all three of which are Raman and infra-red active.

There is one more rule and one more rule-of-thumb that we need to know before we can start to assign vibrational spectra of molecules. The 'rule' is that *polarised* bands in a Raman spectrum can only come from A_1 vibrations. Hence for H_2O, the A_1 vibrations appear in the Raman spectrum as polarised bands, whereas the B_2 vibration appears as a depolarised band. The rule-of-thumb is that stretching vibrations usually appear at higher frequency that bending vibrations.

---

* 'Tensor' is a mathematical term used to denote a matrix with vectors as elements; this definition is not important in our discussion about group theory.

N.B. Any molecule which possesses a centre of symmetry, *i*, is subject to something known as the *mutual exclusion rule*. This rule states that any vibrational band in such a molecule cannot be *both* IR-active and Raman-active.

We are now in a position to take any fairly simple molecule, and decide the number and symmetries of particular vibrations that we expect to see in both the infra-red and Raman spectra of the molecule. Moreover, using projection operators, we can 'see' what the vibrations of the molecule actually look like. Try the following problem:

Tetrachloromethane, CCl_4, has the following vibrational spectroscopy data:

Infra-red spectrum /cm^{-1}	Raman spectrum /cm^{-1}
–	218 depolarised
305	314 depolarised
–	458 polarised
768	762 depolarised

Assign the spectra.

(Hint: first assign the point group, then calculate Γ_{3N}. Using internal coordinates calculate Γ_{str} and Γ_{bend}. To calculate Γ_{bend}, use the difference and internal coordinates methods. Determine which vibrational bands are infra-red and/or Raman active, and then assign the spectra.)

Assign the point group

Calculate Γ_{3N}

$E = 3$ $i = -3$

$\sigma = 1$ $C_n = 1 + 2\cos(360/n)°$ $S_n = -1 + 2\cos(360/n)°$

Using internal coordinates, calculate Γ_{str}.

Using internal coordinates, calculate Γ_{bend}. (The displacement coordinates are the six angles in the molecule. Great care is needed here in figuring out how many of the angles are unshifted on the symmetry operations.)

Determine which bands are Raman and/or infra-red active

Assign the spectra

Answer

The point group is $\mathbf{T}_d$

To calculate Γ_{3N} we can use the following table.

$\mathbf{T}_d$	E	$8C_3$	$3C_2$	$6S_4$	$6\sigma_d$
Number of unshifted atoms	5	2	1	1	3
Contribution to trace per unshifted atom	3	0	−1	−1	1
Γ_{3N}	15	0	−1	−1	3

Using the reduction formula

$$A_1 = \frac{1}{24}\big((15\times1\times1) + (0\times1\times8) + (-1\times1\times3) + (-1\times1\times6) + (3\times1\times6)\big) = 1$$

$$A_2 = \frac{1}{24}\big((15\times1\times1) + (0\times1\times8) + (-1\times1\times3) + (-1\times-1\times6) + (3\times-1\times6)\big) = 0$$

$$E = \frac{1}{24}\big((15\times2\times1) + (0\times-1\times8) + (-1\times2\times3) + (-1\times0\times6) + (3\times0\times6)\big) = 1$$

$$T_1 = \frac{1}{24}\big((15\times3\times1) + (0\times0\times8) + (-1\times-1\times3) + (-1\times1\times6) + (3\times-1\times6)\big) = 1$$

$$T_2 = \frac{1}{24}\big((15\times3\times1) + (0\times0\times8) + (-1\times-1\times3) + (-1\times-1\times6) + (3\times1\times6)\big) = 3$$

Gives $\Gamma_{3N} = A_1 + E + T_1 + 3T_2$

(Since E is doubly degenerate, and T is triply degenerate, Γ_{3N} corresponds to 15 different 'states', which we would expect, since we have five atoms, each with three degrees of freedom.)

From the character table

$$\Gamma_{T+R} = T_1 + T_2$$

(Remember that the T states are triply degenerate, and each single 'T' contains three energy states. Therefore, T_1 accounts for the three rotations.)

Therefore:

$$\Gamma_{vib} = \Gamma_{3N} - \Gamma_{T+R}$$

$$\Gamma_{vib} = A_1 + E + 2T_2$$

Using internal coordinates, we can now calculate Γ_{str}. The bond displacement coordinates are shown in the figure below:

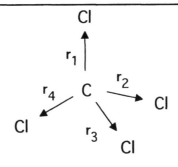

$\mathbf{T}_d$	E	$8C_3$	$3C_2$	$6S_4$	$6\sigma_d$
Number of unshifted	4	1	0	0	2
Γ_{str}	4	1	0	0	2

Using the reduction formula

$$A_1 = \frac{1}{24}\big((4\times1\times1) + (1\times1\times8) + (0\times1\times3) + (0\times1\times6) + (2\times1\times6)\big) = 1$$

$$A_2 = \frac{1}{24}\big((4\times1\times1) + (1\times1\times8) + (0\times1\times3) + (0\times-1\times6) + (2\times-1\times6)\big) = 0$$

$$E = \frac{1}{24}\big((4\times2\times1) + (1\times-1\times8) + (0\times2\times3) + (0\times0\times6) + (2\times0\times6)\big) = 0$$

$$T_1 = \frac{1}{24}\big((4\times3\times1) + (1\times0\times8) + (0\times-1\times3) + (0\times1\times6) + (2\times-1\times6)\big) = 0$$

$$T_2 = \frac{1}{24}\big((4\times3\times1) + (1\times0\times8) + (0\times-1\times3) + (0\times-1\times6) + (2\times1\times6)\big) = 1$$

Gives $\Gamma_{str} = A_1 + T_2$

We can calculate Γ_{bend} from the difference of Γ_{str} and Γ_{vib} to give

$$\Gamma_{bend} = E + T_2$$

Using internal coordinates, we can also calculate Γ_{bend}. The internal coordinates are simply the angles of the molecule, as shown in the figure below.

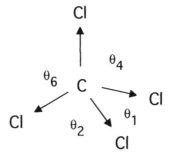

θ_3 and θ_5 omitted for clarity

$\mathbf{T}_d$	E	$8C_3$	$3C_2$	$6S_4$	$6\sigma_d$
Number of unshifted angles	6	0	2	0	2
Γ_{bend}	6	0	2	0	2

Using the reduction formula

$$A_1 = \frac{1}{24}\left((6\times1\times1) + (0\times1\times8) + (2\times1\times3) + (0\times1\times6) + (2\times1\times6)\right) = 1$$

$$A_2 = \frac{1}{24}\left((6\times1\times1) + (0\times1\times8) + (2\times1\times3) + (0\times-1\times6) + (2\times-1\times6)\right) = 0$$

$$E = \frac{1}{24}\left((6\times2\times1) + (0\times-1\times8) + (2\times2\times3) + (0\times0\times6) + (2\times0\times6)\right) = 1$$

$$T_1 = \frac{1}{24}\left((6\times3\times1) + (0\times0\times8) + (2\times-1\times3) + (0\times1\times6) + (2\times-1\times6)\right) = 0$$

$$T_2 = \frac{1}{24}\left((6\times3\times1) + (0\times0\times8) + (2\times-1\times3) + (0\times-1\times6) + (2\times1\times6)\right) = 1$$

Gives $\Gamma_{bend} = A_1 + E + T_2$

This answer contains an extra A_1 from that calculated by difference. The extra A_1 here is called a redundant coordinate, and arises because of a mathematical reason (the angles for the basis for Γ_{bend} are not orthogonal). The redundant coordinate can be safely ignored.

Therefore: $\Gamma_{str} = A_1 + T_2$
$\Gamma_{bend} = E + T_2$

Using the direct product rules we can now decide which of the bands are infra-red active. From the T_d character table, the dipole of the molecule is represented by T_2.

For A_1 $A_1 \times T_2 \times A_1 = T_2$ (not IR active)
For T_2 $A_1 \times T_2 \times T_2 = A_1 + E + T_1 + T_2$ (IR active)
For E $A_1 \times T_2 \times E = T_1 + T_2$ (not IR active)

Hence, we expect two IR bands: one T_2 stretch, and one T_2 bend.

From the T_d character table, the polarisability tensor is represented by A_1, E and T_2.

For A_1 $A_1 \times A_1 \times A_1 = A_1$ (Raman active)
$A_1 \times E \times A_1 = E$
$A_1 \times T_2 \times A_1 = T_2$

For T_2 $A_1 \times A_1 \times T_2 = T_2$
$A_1 \times E \times T_2 = T_1 + T_2$
$A_1 \times T_2 \times T_2 = A_1 + E + T_2$ (Raman active)

For E $A_1 \times A_1 \times E = E$
$A_1 \times E \times E = A_1 + E$ (Raman active)
$A_1 \times T_2 \times E = T_1 + T_2$

Hence, we expect all four bands (two stretches and two bends) to be Raman active.

Since we know that the polarised band in the Raman is the A_1 stretch, and we know that the frequency of a bending vibration is usually greater than that of a stretching vibration we can assign the spectra as follows:

Infra-red spectrum /cm^{-1}	Raman spectrum /cm^{-1}	Assignment
–	218 depolarised	E bend
305	314 depolarised	T_2 bend
–	458 polarised	A_1 stretch
768	762 depolarised	T_2 stretch

6.6   Summary

This section has shown how group theory can be used powerfully in the assignment of vibrational spectra. In particular we have seen how to generate a reducible representation using displacement coordinates as a basis. These bases have included the full displacements of the atoms, or the displacements of the atoms along bonds or about angles. The last two bases are known as internal coordinates, and provide a useful short-cut to determining the possible vibrations of a molecule.

The key points of the section are:

*   Molecular vibrations can be represented with displacement coordinates.

*   The displacement coordinates form a basis for a reducible representation.

*   Internal coordinates can be used as a basis.

*   The reducible representations can be reduced to give the number of possible vibrational modes.

*   The IR and Raman activity of such modes can be determined using direct products.

SECTION 7

Group theory in action: molecular orbitals

Group theory in action: molecular orbitals

7.1   Some general points

Having seen that group theory is very powerful in analysing molecular vibrations, we now turn our attention to molecular orbitals. Before we start this section, it is worth going over some general points about molecular orbitals (MOs). Hopefully, this should be revision. The concept of an MO comes right out of MO theory. The theory was developed to describe the bonding known to exist in some simple molecules. For example, MO theory was the first theory that could explain the paramagnetism of O_2. An MO is constructed from the linear combination of atomic orbitals (LCAO). So, for example, consider two atoms coming together, one with an atomic orbital denoted by χ_a and the other atom with an atomic orbital denoted by χ_b. An MO (ψ) is formed from the linear combination of these two atomic orbitals:

$$\psi = c_a\chi_a + c_b\chi_b$$

where c_a and c_b are coefficients.

Can we determine the possible coefficients? Since the MO must be a solution to the Schrödinger equation for that system, the *combination* of the two atomic orbitals must form a basis for the point group of the molecule.* Hence, we can use group theory to restrict greatly the possible MOs that can be formed. In fact, group theory can tell us the coefficients of the atomic orbitals in the LCAO. In essence, we use group theory to give us a symmetry adapted linear combination (SALC) of the atomic orbitals.

It is easiest to see how this works by looking at some examples. The examples are divided into two classes, one class where the molecule contains a 'central' atom, and the other where there is no 'central' atom.

7.2   Molecules with a 'central' atom

First of all, what do we mean by a 'central' atom. A 'central' atom is an atom which is not shifted in space by any symmetry operations of the point group of the molecule, i.e. the atom lies on the 'point' of the group.

Which is the 'central' atom in the CH_4 molecule? Confirm that the atom lies on the point of the group.

Answer

The C atom is the central atom, and its position is not changed by any of the symmetry operations of the T_d point group.

---

* From a mathematical point of view, we have to make use of the fact that the Hamiltonian operator is a *linear* operator. In other words. $aH\psi_a + bH\psi_b = H(a\psi_a + b\psi_b)$.

In many cases it is easy to spot the 'central' atom, and it is not necessary to check whether it is unshifted in position by the symmetry operations of the point group.

What is special about the central' atom? As we saw in Section 5, the irreducible representations of the atomic orbitals on this atom can be read directly from the character table. Try the following:

In the following, what irreducible representations do the atomic orbitals belong to?

p_z orbital on the N atom in NH_3
s orbital on the C atom in CH_4
d_{xy} orbital on the Au atom in $AuBr_4^-$
d_{z^2} orbital on the P atom in PCl_5

Answer

NH_3 is point group C_{3v}. The p_z orbital has A_1 symmetry
CH_4 is point group T_d. The s orbital has A_1 symmetry
$AuBr_4^-$ is point group D_{4h}. The d_{xy} orbital has B_{2g} symmetry
PCl_5 is point group D_{3h}. The d_{z^2} orbital has A_1' symmetry

---

What irreducible representations do the valence atomic orbitals on the oxygen atom in H_2O belong to?

Answer

H_2O is point group C_{2v}. The valence atomic orbitals on O are $2s$, $2p_x$, $2p_y$ and $2p_z$.

From the C_{2v} character table:

$2s$ has A_1 symmetry
$2p_x$ has B_1 symmetry
$2p_y$ has B_2 symmetry
$2p_z$ has A_1 symmetry

Notice that in MO theory we only have to consider the valence orbitals when it comes to generating an MO. This is because the non-valence orbitals are either fully occupied or fully unoccupied by electrons, and any MOs generated from these atomic orbitals are, to a good approximation, as much anti-bonding as bonding, and are not important in determining the stability of the molecule.

Knowing the symmetry of the orbitals on the 'central' atom, we now have to work out how these orbitals can combine with orbitals on the other atoms. To be able to combine these orbitals we must know their symmetries, since one of the key rules in MO theory is that *only atomic orbitals of the same symmetry can combine to give bonding and antibonding MOs*. Take the water molecule for example; what are the symmetries of the orbitals on the hydrogen atoms. As we saw in Section 5, we cannot consider the two hydrogen orbitals separately, we must consider combinations of the orbitals; this leads us to ask a slightly different question. What are the symmetries of the allowable *combinations* of hydrogen atoms?

Well, this is an easy question to answer, all we need to do is generate a reducible representation of the hydrogen 1s orbitals and reduce it. Reduction will give us the symmetries of the SALCs that we are looking for. An example will hopefully make this clear.

Step 1 is to generate a reducible representation of the hydrogen atom 1s orbitals. The procedure is similar to the last section. We are required to figure out the number of unshifted hydrogen atoms on each operation, and then multiply these numbers with the contribution to the character (or trace) of the representation. For s orbitals the contribution to the character is always 1.

Hence we can draw up the following table.

C_{2v}	E	C_2	σ_v	σ_v'
No. of unshifted H atoms 2	0	0	2	
Γ_H 2	0	0	2	

Use the reduction formula to reduce this representation.

Answer

Using the same procedure as on pages 86 and 87.

$\Gamma_H = A_1 + B_2$

Hence the SALCs of the two hydrogen orbitals have symmetries A_1 and B_2. We know from Section 5 that the un-normalised forms of $A_1 = 1s_{H(1)} + 1s_{H(2)}$ and $B_1 = 1s_{H(1)} - 1s_{H(2)}$.

We are now in a position to draw up a qualitative molecular-orbital energy level diagram for the water molecule, using the rule that only orbitals of the same symmetry can combine. It is assumed at this point that the reader is familiar with the principles of generating bonding, non-bonding and anti-bonding MOs. It is important also that the reader realises one of the limitations of qualitative group theory at this point. In generating an MO energy level diagram, group theory can tell us which orbitals to combine, but it does not tell us the energies of the new MOs. We cannot calculate the energies of the MOs without having to embark on some fairly involved mathematics, and we are left with making educated guesses. This necessarily requires some knowledge of what the MO looks like. So, for instance, it is usually a safe bet that MOs involving s orbitals will give lower energy bonding and higher energy anti-bonding levels, than the same MOs built up from p orbitals. Using these rules-of-thumb gets better with practice.

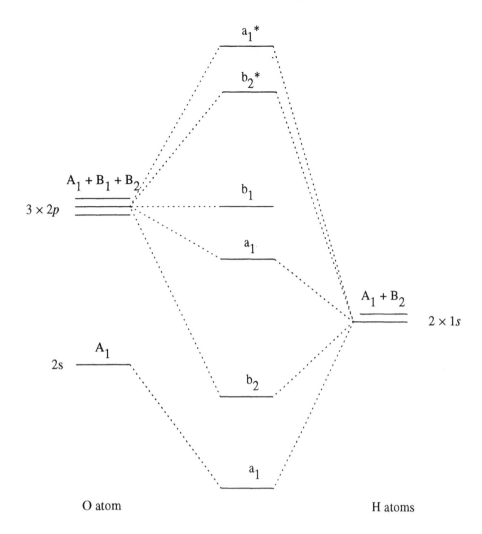

O atom   H atoms

Note that for the sake of clarity that not all of the correlation lines (i.e. dotted lines on diagram) have been drawn in.

Our MO diagram of H_2O shows six MOs with two bonding, two non-bonding and two antibonding levels. Of note is the B_1 (p_x) orbital on the oxygen atom, which has no symmetry equivalent in the H atom SALCs. The b_1 level, therefore, remains non-bonding. Also of note is the combination of three A_1 levels to give bonding, near non-bonding and anti-bonding levels.

There are eight electrons to put in this MO diagram (six valence electrons on the O atom and two electrons on the two H atoms), which gives a net sum of two occupied bonding levels (i.e. two bonds) and two approximately non-bonding levels (a_1 and b_1). The two non-bonding levels correspond to the two lone pairs on the O atom.

We can now use the projection operator method to determine the actual form of the SALCs. Taking the hydrogen atom 1s combination, we can use one of the orbitals as a generating function, as follows.

C_{2v}	E	C_2	σ_v	σ_v'
$1s_{H(1)}$	$1s_{H(1)}$	$1s_{H(2)}$	$1s_{H(2)}$	$1s_{H(1)}$
A_1	$1 \times 1s_{H(1)}$	$1 \times 1s_{H(2)}$	$1 \times 1s_{H(2)}$	$1 \times 1s_{H(1)}$
B_2	$1 \times 1s_{H(1)}$	$-1 \times 1s_{H(2)}$	$-1 \times 1s_{H(2)}$	$1 \times 1s_{H(1)}$

Therefore: A_1 is equivalent to $1s_{H(1)} + 1s_{H(2)}$ (again, we leave these as un-normalised)

B_2 is equivalent to $1s_{H(1)} - 1s_{H(2)}$

In sketch form, these appear as (a negative sign in front of the orbital means that it has the opposite phase)

So, using group theory we can determine the SALCs of orbitals on atoms which do not lie on the point of the group, and then combine these SALCs with the atomic orbitals of the 'central' atom to give us a qualitative MO energy level diagram. This is an extremely useful means of analysing the σ-bonding in molecules.

We can apply exactly the same procedure to far more complicated molecules, where there is a central atom. Note however, we have restricted ourselves to generating SALCs of *s* orbitals only; we have not considered *p* orbitals on the peripheral atoms. We have not treated the possibility of *p* orbitals on the peripheral atoms interacting with *p* orbitals on the central atom, in other words π-bonding. Of course, π-bonding cannot occur with hydrogen atoms, since they do not have *p* orbitals as valence orbitals, but it may well occur with non-hydrogen atoms. We will not consider π-bonding in our discussion about molecules with a 'central' atom, but we will consider π-bonding later in the section when we look at the bonding in polyenes. In this part, we will restrict the examples to ones where only σ-bonding is present.

Try the following three examples, which steadily get more complex.

Sketch a qualitative MO energy level diagram for NH_3. (Hint: first assign the point group, figure out the symmetries of the valence orbitals on the nitrogen atom, then determine the SALCs for the H atoms by generating a reducible representation and reducing it, and finally sketch the MO diagram.)

Answer

The point group of the molecule is C_{3v}

From the C_{3v} character table, the atomic orbitals on the nitrogen atom have the following symmetries:

$2s$ has A_1 symmetry
$2p_x$ and $2p_y$ have E symmetry (note that the E representation is doubly degenerate and accounts for both orbitals).
$2p_z$ has A_1 symmetry

Generate a reducible representation of the three hydrogen atoms as follows:

C_{3v}	E	$2C_3$	$3\sigma_v$
No. of unshifted H atoms	3	0	1
Γ_H	3	0	1

Using the reduction formula

$$A_1 = \frac{1}{6}\left((3\times1\times1) + (0\times1\times2) + (1\times1\times3)\right) = 1$$

$$A_2 = \frac{1}{6}\left((3\times1\times1) + (0\times1\times2) + (-1\times1\times3)\right) = 0 \qquad \text{which gives } \Gamma_H = A_1 + E$$

$$E = \frac{1}{6}\left((3\times2\times1) + (0\times-1\times2) + (1\times0\times3)\right) = 1$$

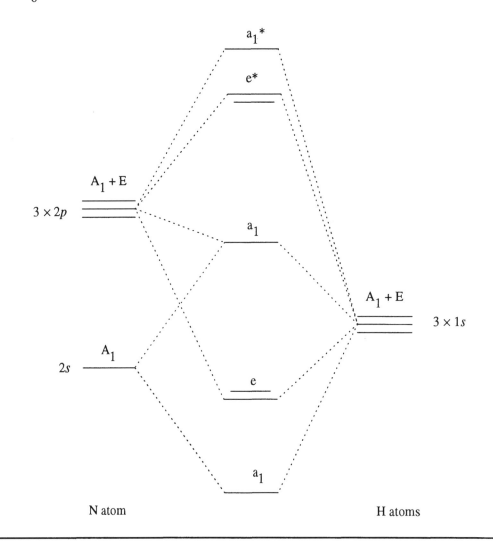

N atom   H atoms

Sketch a qualitative MO energy level diagram for CH_4. (Use the same procedures as for the NH_3 example.)

Answer

The point group is $\mathbf{T}_d$.

From the $\mathbf{T}_d$ character table, the C atom symmetries are: $2s$ is A_1, $2p_x$ $2p_y$ and $2p_z$ are T_2.

Generating the SALCs for the four H atoms

$\mathbf{T}_d$	E	$8C_3$	$3C_2$	$6S_4$	$6\sigma_d$
No. of unshifted H atoms	4	1	0	0	2
Γ_H	4	1	0	0	2

Using the reduction formula (ignore the C_2 and S_4 contributions as they are zero in all cases).

$$A_1 = \frac{1}{24}\big((4\times1\times1) + (1\times1\times8) + (2\times1\times6)\big) = 1$$

$$A_2 = \frac{1}{24}\big((4\times1\times1) + (1\times1\times8) + (2\times-1\times6)\big) = 0$$

$$E = \frac{1}{24}\big((4\times2\times1) + (1\times-1\times8) + (2\times0\times6)\big) = 0 \qquad \text{which gives } \Gamma_H = A_1 + T_2$$

$$T_1 = \frac{1}{24}\big((4\times3\times1) + (1\times0\times8) + (2\times-1\times6)\big) = 0$$

$$T_2 = \frac{1}{24}\big((4\times3\times1) + (1\times0\times8) + (2\times1\times6)\big) = 1$$

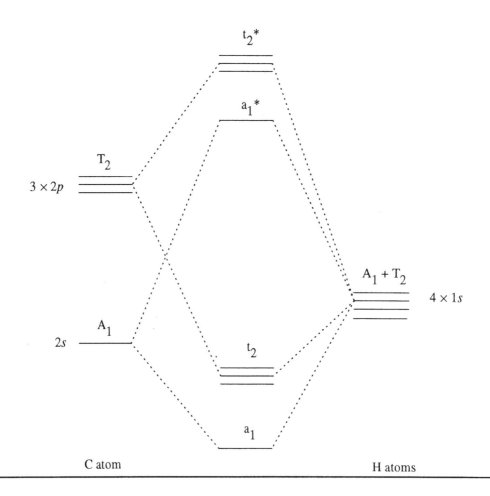

Sketch a qualitative MO energy level diagram for $CoCl_6^{3-}$. (This is quite difficult and do not worry if you do not manage to get it all correct. Hint: we must now consider the $3d$, $4s$ and $4p$ valence orbitals on the cobalt atom. Ignore any p orbitals on the Cl atoms, and assume that the valence orbitals are s orbitals. First figure out the point group of the molecule, then determine the symmetries of the valence atomic orbitals on the Co atom. Generate a reducible representation of the s orbitals on the Cl atoms, reduce the representation, and then sketch the MO diagram combining orbitals of the same symmetry. To help, the energy levels of the atoms are drawn at the bottom of page 124.)

4p ═══

4s ───

3d ═══

Co atom   Cl atoms

Answer

The point group is O_h

From the O_h character table, the symmetries of the Co atomic orbitals are:-

$4s$ is A_{1g},

$4p$ are T_{1u},

$3d$ are T_{2g} and E_g.

Generate a reducible representation for the Cl atom s orbitals. Some of these symmetry operations may be difficult to 'see'. Some help can be found in Section 2. It is important to realise that there are three principal C_4 axes in the molecule, therefore there are three σ_h. The σ_d planes, parallel with the principal axes, bisect the Cl–Co–Cl angles. There are two sets of C_2 axes. One set is coaxial with the principal axes, the other set bisects the Cl–Co–Cl angles.

O_h	E	$8C_3$	$6C_2$	$6C_4$	$3C_2(C_4{}^2)$	i	$6S_4$	$8S_6$	$3\sigma_h$	$6\sigma_d$
Number of unshifted Cl	6	0	0	2	2	0	0	0	4	2
Γ_{Cl}	6	0	0	2	2	0	0	0	4	2

Using the reduction formula, again ignoring symmetry operations with 0 in reducible representation.

$$A_{1g} = \frac{1}{48}\left((6\times1\times1) + (2\times1\times6) + (2\times1\times3) + (4\times1\times3) + (2\times1\times6)\right) = 1$$

$$A_{2g} = \frac{1}{48}\left((6\times1\times1) + (2\times-1\times6) + (2\times1\times3) + (4\times1\times3) + (2\times-1\times6)\right) = 0$$

$$E_g = \frac{1}{48}\left((6\times2\times1) + (2\times0\times6) + (2\times2\times3) + (4\times2\times3) + (2\times0\times6)\right) = 1$$

$$T_{1g} = \frac{1}{48}\left((6\times3\times1) + (2\times1\times6) + (2\times-1\times3) + (4\times-1\times3) + (2\times-1\times6)\right) = 0$$

$$T_{2g} = \frac{1}{48}\left((6\times3\times1) + (2\times-1\times6) + (2\times-1\times3) + (4\times-1\times3) + (2\times1\times6)\right) = 0$$

$$A_{1u} = \frac{1}{48}\left((6\times1\times1) + (2\times1\times6) + (2\times1\times3) + (4\times-1\times3) + (2\times-1\times6)\right) = 0$$

$$A_{2u} = \frac{1}{48}\left((6\times1\times1) + (2\times-1\times6) + (2\times1\times3) + (4\times-1\times3) + (2\times1\times6)\right) = 0$$

$$E_u = \frac{1}{48}\left((6\times2\times1) + (2\times0\times6) + (2\times2\times3) + (4\times-2\times3) + (2\times0\times6)\right) = 0$$

$$T_{1u} = \frac{1}{48}\left((6\times3\times1) + (2\times1\times6) + (2\times-1\times3) + (4\times1\times3) + (2\times1\times6)\right) = 1$$

$$T_{2g} = \frac{1}{48}\left((6\times3\times1) + (2\times-1\times6) + (2\times-1\times3) + (4\times1\times3) + (2\times-1\times6)\right) = 0$$

This gives $\Gamma_{Cl} = A_{1g} + E_g + T_{1u}$

We can now generate a MO energy level diagram. Note that the T_{2g} levels on the d orbitals of the metal atom have no ligand equivalent, therefore, they remain non-bonding.

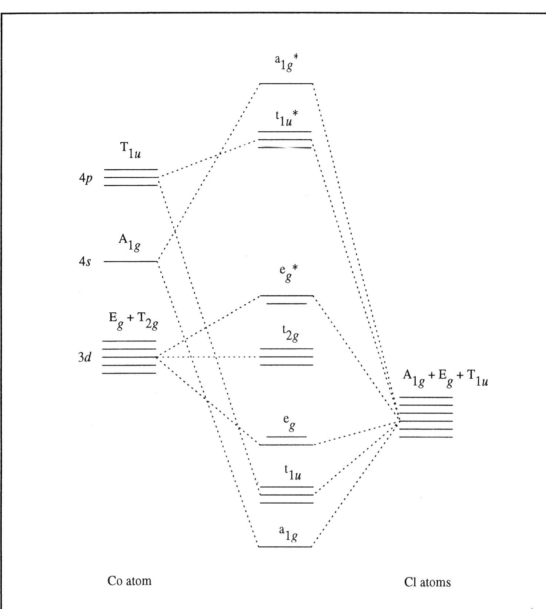

Co atom   Cl atoms

It is worth noting at this point that we have generated a set of MOs which describe the bonding between the metal ion and the ligands. In particular the bonding of the d orbitals is described by this diagram. You may notice that there is a non-bonding triply degenerate t_{2g} level adjacent to an anti-bonding doubly degenerate e_g level. This is exactly the pattern of d orbital splittings that we predict from crystal field theory for an octahedral metal complex.

7.3   Molecules with π-bonding and without a 'central' atom

Having seen that we can use group theory to generate really fairly complicated MO energy level diagrams, we now turn our attention to molecules which do not have a 'central' atom, and are also capable of π-bonding. We are going to restrict our discussion to polyenes, e.g. butadiene and benzene, since group theory is particularly useful in determining the possible π-bonding energy levels in these molecules. Furthermore, the richness of these molecules' organic chemistry means that they have been studied extensively in terms of their frontier orbitals. These studies have shown that an understanding of the possible energy levels can help rationalise the molecules' reactivities. We will not cover any of these reactivity studies here (perhaps the most famous studies are the Woodward–Hoffman rules for polyene cyclisation reactions), but the results from our discussion provide a foundation for understanding frontier orbitals.

Let us start with an example. (*E*)-butadiene, see figure, has two double bonds. By sketching resonance structures we predict that the molecule will contain a delocalised π system across the whole molecule. Can we use group theory to tell us about this π system?

(*E*)-butadiene

We start by simplifying the problem. All we are concerned with is the π system, which is formed from the overlap of four *p* orbitals perpendicular to the plane of the molecule. We are not concerned with any of the other bonds since these are not valence energy levels. Hence, all we need to consider is how the four *p* orbitals perpendicular to the plane overlap with each other. (N.B. We are only able to do this since we know that these *p* orbitals are not involved in any other type of bonding, if they were then we could not make this simplification.)

How, then, do these four orbitals combine? Group theory tells us that the linear combination of the four orbitals must form a basis for the point group of the molecule. Hence, we should be able to generate a reducible representation from the orbitals.

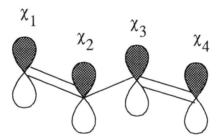

π-bonding *p* orbitals in (*E*)-butadiene

Assign the point group of (*E*)-butadiene? (assume a rigid planar structure)

Answer

The point group is C_{2h}

Generate a reducible representation using as a basis the four p orbitals which are perpendicular to the plane of the (E)-butadiene molecule. (N.B. Unlike s-orbitals the contribution to the character of each p orbital per unshifted atom is *not* always 1. In this special case of looking at p orbitals which are parallel to the principal axis of the molecule, we can say that if a p orbital is not inverted by the symmetry operation then the contribution is 1, but if a p-orbital is inverted by the symmetry operation then the contribution is −1 per unshifted atom.)

Reduce the representation using the reduction formula.

Answer

C_{2h}	E	C_2	i	σ_h
No. unshifted atoms	4	0	0	4
Contribution per unshifted atom	1	1	−1	−1
Γ_π	4	0	0	−4

Reducing the representation (ignoring the symmetry operations where Γ_π is zero):

$$A_g = \frac{1}{4}\left((4\times1\times1) + (-4\times1\times1)\right) = 0$$

$$B_g = \frac{1}{4}\left((4\times1\times1) + (-4\times-1\times1)\right) = 2$$

$$A_u = \frac{1}{4}\left((4\times1\times1) + (-4\times-1\times1)\right) = 2$$

$$B_u = \frac{1}{4}\left((4\times1\times1) + (-4\times1\times1)\right) = 0$$

Hence $\Gamma_\pi = 2a_u + 2b_g$

(Notice that we have used lower case letters, as these now represent MOs.)

What do the SALCs of the $2a_u$ and $2b_g$ representations look like? To answer this we can use the projection operator method. For generating functions we need to use $\chi_1 + \chi_2$ and $\chi_1 - \chi_2$. The choice of these as generating functions is beyond this text.

Using the projection operator method to generate SALCs for the π-bonding in butadiene, using $\chi_1 + \chi_2$ and $\chi_1 - \chi_2$ as generating functions. Use each generating function for both a_u and b_g representations. Sketch the result.

Answer

C_{2h}	E	C_2	i	σ_h
$\chi_1 + \chi_2$	$\chi_1 + \chi_2$	$\chi_3 + \chi_4$	$-\chi_3 - \chi_4$	$-\chi_1 - \chi_2$
a_u	$\chi_1 + \chi_2$	$\chi_3 + \chi_4$	$\chi_3 + \chi_4$	$\chi_1 + \chi_2$
b_g	$\chi_1 + \chi_2$	$-\chi_3 - \chi_4$	$-\chi_3 - \chi_4$	$\chi_1 + \chi_2$

Which corresponds to
a_u is $\chi_1 + \chi_2 + \chi_3 + \chi_4$
b_g is $\chi_1 + \chi_2 - \chi_3 - \chi_4$

	E	C_2	i	σ_h
$\chi_1 - \chi_2$	$\chi_1 - \chi_2$	$-\chi_3 + \chi_4$	$\chi_3 - \chi_4$	$-\chi_1 + \chi_2$
a_u	$\chi_1 - \chi_2$	$-\chi_3 + \chi_4$	$-\chi_3 + \chi_4$	$\chi_1 - \chi_2$
b_g	$\chi_1 - \chi_2$	$\chi_3 - \chi_4$	$\chi_3 - \chi_4$	$\chi_1 - \chi_2$

Which corresponds to:
a_u is $\chi_1 - \chi_2 - \chi_3 + \chi_4$
b_g is $\chi_1 - \chi_2 + \chi_3 - \chi_4$

In sketch form:

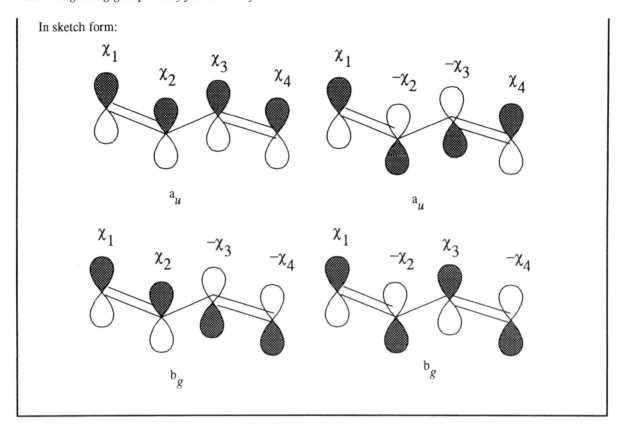

From the SALCs pictured above, we can envisage MOs that can be generated by these particular combinations of *p* orbitals. Moreover, we can now use these SALCs to come-up with a qualitative MO energy level diagram for the π-bonding in (*E*)-butadiene. This can be done by counting the number of nodes (i.e. areas of zero electron density) which are perpendicular to the plane of the molecule.

On the diagram above indicate the nodes with a dotted line which are perpendicular to the plane of the molecule.

Answer

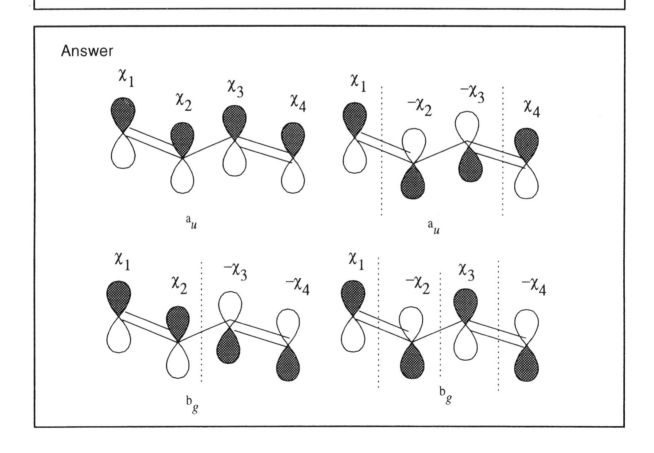

Using the rule-of-thumb that more nodes mean a higher energy (i.e. more anti-bonding orbital) we can draw up a qualitative MO energy level diagram as follows:

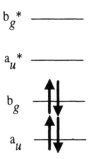

The * represents an MO which is more anti-bonding than it is bonding. With four electrons to put in we see that the π-system is fully bonding.

Hence, we see that group theory is very useful in helping to determine the energy levels in the π system of conjugated dienes. Can we take this to more complicated molecules? Try the following exercise.

Using the p orbitals (those parallel to the principal axis) as a basis, determine the SALCs for the π-bonding levels in cyclobutadiene (see figure). When generating the SALCs, use χ_1 as a generating function, and use χ_2 as the generating function for the second e_g SALC. (Hint: use the same procedure as above.)

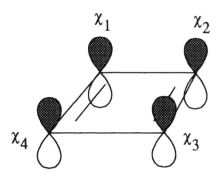

Assign the point group.

Generate the reducible representation.

Reduce the representation.

Use the projection operator method to generate SALCs (remember to use χ_1 as a generating function, and use χ_2 as the generating function for the second e_g SALC). Remember also to use the expanded form of the character table.

Sketch the SALCs.

Sketch the MO energy level diagram.

Fill in the electrons in the MO energy level diagram, and say whether you expect isolated cyclobutadiene to be a stable or unstable molecule.

Answer

The point group is D_{4h}.

Using the p orbitals as a basis:

D_{4h}	E	$2C_4$	C_2	$2C_2$	$2C_2''$	i	$2S_4$	σ_h	$2\sigma_v$	$2\sigma_d$
No. of unshifted atoms	4	0	0	2	0	0	0	4	2	0
Contribution per unshifted atom	1	1	1	−1	−1	−1	−1	−1	1	1
Γ_π	4	0	0	−2	0	0	0	−4	2	0

Using the reduction formula:

$$a_{1g} = \frac{1}{16}\left(\begin{array}{l}(4\times1\times1) + (0\times1\times2) + (0\times1\times1) + (-2\times1\times2) + (0\times1\times2) \\ + (0\times1\times1) + (0\times1\times2) + (-4\times1\times1) + (2\times1\times2) + (0\times1\times2)\end{array}\right) = 0$$

$$a_{2g} = \frac{1}{16}\left(\begin{array}{l}(4\times1\times1) + (0\times1\times2) + (0\times1\times1) + (-2\times-1\times2) + (0\times-1\times2) \\ + (0\times1\times1) + (0\times1\times2) + (-4\times1\times1) + (2\times-1\times2) + (0\times-1\times2)\end{array}\right) = 0$$

$$b_{1g} = \frac{1}{16}\left(\begin{array}{l}(4\times1\times1) + (0\times-1\times2) + (0\times1\times1) + (-2\times1\times2) + (0\times-1\times2) \\ + (0\times1\times1) + (0\times-1\times2) + (-4\times1\times1) + (2\times1\times2) + (0\times-1\times2)\end{array}\right) = 0$$

$$b_{2g} = \frac{1}{16}\left(\begin{array}{l}(4\times1\times1) + (0\times-1\times2) + (0\times1\times1) + (-2\times-1\times2) + (0\times1\times2) \\ + (0\times1\times1) + (0\times-1\times2) + (-4\times1\times1) + (2\times-1\times2) + (0\times1\times2)\end{array}\right) = 0$$

$$e_g = \frac{1}{16}\left(\begin{array}{l}(4\times2\times1) + (0\times0\times2) + (0\times-2\times1) + (-2\times0\times2) + (0\times0\times2) \\ + (0\times2\times1) + (0\times0\times2) + (-4\times-2\times1) + (2\times0\times2) + (0\times0\times2)\end{array}\right) = 1$$

$$a_{1u} = \frac{1}{16}\left(\begin{array}{l}(4\times1\times1) + (0\times1\times2) + (0\times1\times1) + (-2\times1\times2) + (0\times1\times2) \\ + (0\times-1\times1) + (0\times-1\times2) + (-4\times-1\times1) + (2\times-1\times2) + (0\times-1\times2)\end{array}\right) = 0$$

$$a_{2u} = \frac{1}{16}\begin{pmatrix} (4\times1\times1) + (0\times1\times2) + (0\times1\times1) + (-2\times-1\times2) + (0\times-1\times2) \\ + (0\times-1\times1) + (0\times-1\times2) + (-4\times-1\times1) + (2\times1\times2) + (0\times1\times2) \end{pmatrix} = 1$$

$$b_{1u} = \frac{1}{16}\begin{pmatrix} (4\times1\times1) + (0\times-1\times2) + (0\times1\times1) + (-2\times1\times2) + (0\times-1\times2) \\ + (0\times-1\times1) + (0\times1\times2) + (-4\times-1\times1) + (2\times-1\times2) + (0\times1\times2) \end{pmatrix} = 0$$

$$b_{2u} = \frac{1}{16}\begin{pmatrix} (4\times1\times1) + (0\times-1\times2) + (0\times1\times1) + (-2\times-1\times2) + (0\times1\times2) \\ + (0\times-1\times1) + (0\times1\times2) + (-4\times-1\times1) + (2\times1\times2) + (0\times-1\times2) \end{pmatrix} = 1$$

$$e_u = \frac{1}{16}\begin{pmatrix} (4\times2\times1) + (0\times0\times2) + (0\times-2\times1) + (-2\times0\times2) + (0\times0\times2) \\ + (0\times-2\times1) + (0\times0\times2) + (-4\times2\times1) + (2\times0\times2) + (0\times0\times2) \end{pmatrix} = 0$$

Which gives $\Gamma_\pi = a_{2u} + b_{2u} + e_g$.

We can now use the projection operator method to generate the SALCs for the a_{2u}, b_{2u} and e_g representations, see below.

If we do this:

$\mathbf{D_{4h}}$	E	$C_4{}^1$	$C_4{}^3$	C_2	$C_2'(x)$	$C_2'(y)$	$C_2''(1)$	$C_2''(2)$	i	$S_4{}^1$	$S_4{}^3$	σ_h	$\sigma_v(x)$	$\sigma_v(y)$	$\sigma_d(1)$	$\sigma_d(2)$
χ_1	χ_1	χ_2	χ_4	χ_3	$-\chi_1$	$-\chi_3$	$-\chi_2$	$-\chi_4$	$-\chi_3$	$-\chi_2$	$-\chi_4$	$-\chi_1$	χ_1	χ_3	χ_4	χ_2
a_{2u}	χ_1	χ_2	χ_4	χ_3	χ_1	χ_3	χ_2	χ_4	χ_3	χ_2	χ_4	χ_1	χ_1	χ_3	χ_4	χ_2
b_{2u}	χ_1	$-\chi_2$	$-\chi_4$	χ_3	χ_1	χ_3	$-\chi_2$	$-\chi_4$	χ_3	$-\chi_2$	$-\chi_4$	χ_1	χ_1	χ_3	$-\chi_4$	$-\chi_2$
e_g	$2\chi_1$	$0\chi_2$	$0\chi_4$	$-2\chi_3$	$0\chi_1$	$0\chi_3$	$0\chi_2$	$0\chi_4$	$-2\chi_3$	$0\chi_2$	$0\chi_4$	$2\chi_1$	$0\chi_1$	$0\chi_3$	$0\chi_4$	$0\chi_2$

which gives:

      a_{2u} is equivalent to: $4\chi_1 + 4\chi_2 + 4\chi_3 + 4\chi_4$, which is equivalent to $\chi_1 + \chi_2 + \chi_3 + \chi_4$
      b_{2u} is equivalent to: $4\chi_1 - 4\chi_2 + 4\chi_3 - 4\chi_4$, which is equivalent to $\chi_1 - \chi_2 + \chi_3 - \chi_4$
      e_g is equivalent to: $4\chi_1 - 4\chi_3$, which is equivalent to $\chi_1 - \chi_3$

For the remaining e_g SALC, use χ_2 as a generating coordinate

D_{4h}	E	$C_4{}^1$	$C_4{}^3$	C_2	$C_2'(x)$	$C_2'(y)$	$C_2''(1)$	$C_2''(2)$	i	$S_4{}^1$	$S_4{}^3$	σ_h	$\sigma_v(x)$	$\sigma_v(y)$	$\sigma_d(1)$	$\sigma_d(2)$
χ_2	χ_2	χ_3	χ_1	χ_4	$-\chi_2$	$-\chi_4$	$-\chi_3$	$-\chi_1$	$-\chi_4$	$-\chi_3$	$-\chi_1$	$-\chi_2$	χ_2	χ_4	χ_1	χ_3
e_g	$2\chi_2$	$0\chi_3$	$0\chi_1$	$-2\chi_4$	$0\chi_2$	$0\chi_4$	$0\chi_3$	$0\chi_1$	$-2\chi_4$	$0\chi_3$	$0\chi_1$	$2\chi_2$	$0\chi_2$	$0\chi_4$	$0\chi_1$	$0\chi_3$

which gives

      e_g is equivalent to: $4\chi_2 - 4\chi_4$, which is equivalent to $\chi_2 - \chi_4$

Hence, our SALCs are

a_{2u}	$\chi_1 + \chi_2 + \chi_3 + \chi_4$
b_{2u}	$\chi_1 - \chi_2 + \chi_3 - \chi_4$
e_g	$\chi_1 - \chi_3$ and $\chi_2 - \chi_4$

In sketch form:

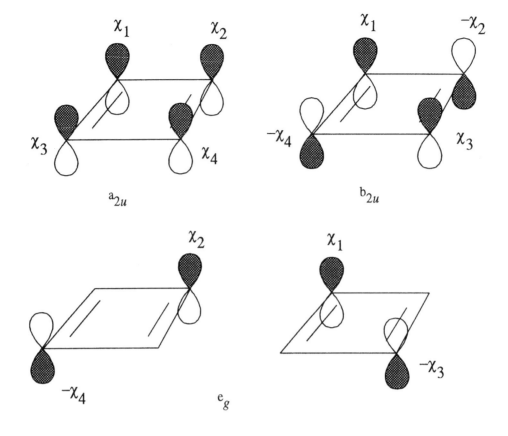

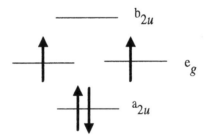

Counting nodes we can sketch a qualitative MO energy level diagram

Notice that if we put in the electrons, we predict that cyclobutadiene is a diradical, which is usually very unstable. Indeed, isolated cyclobutadiene is known to be very unstable as an anti-aromatic molecule.

7.4   Summary

We have seen in this section that group theory can be used to generate linear combinations of atomic orbitals, which represent molecular orbitals. The generated LCAOs are known as *symmetry adapted linear combinations* or SALCs. The SALCs give us an accurate picture of MOs. In particular, we can use the SALCs to generate qualitative MO energy level diagrams, which are useful in predicting the properties of a molecule.

The main points of the section are:

- Atomic orbitals can be used as a basis in group theory.
- SALCs can be generated from atomic orbitals by generating a reducible representation and then reducing it.
- Using the LCAO-MO theory the SALCs represent the MOs of a molecule.
- A qualitative MO energy level diagram can be generated from the SALCs.
- Filling in the electrons in the MO energy level diagram can help us predict the properties of molecules.

APPENDIX 1

Character tables common in chemistry

Selected chemically important character tables

1. The non-axial groups

C_s	E	σ_h		
A'	1	1	x, y, R_z	x^2, y^2, z^2, xy
A"	1	−1	z, R_x, R_y	yz, xz

C_i	E	i		
A_g	1	1	R_x, R_y, R_z	x^2, y^2, z^2, xy
A_u	1	−1	x, y, z	yz, xz

2. The C_2 group

C_2	E	C_2		
A	1	1	z, R_z	x^2, y^2, z^2, xy
B	1	−1	x, y, R_x, R_y	yz, xz

3. The D_n groups

D_2	E	$C_2(z)$	$C_2(y)$	$C_2(x)$		
A	1	1	1	1		x^2, y^2, z^2
B_1	1	1	−1	−1	z, R_z	xy
B_2	1	−1	1	−1	y, R_y	xz
B_3	1	−1	−1	1	x, R_x	yz

D_3	E	$2C_3$	$3C_2$		
A_1	1	1	1		$x^2 + y^2$, z^2
A_2	1	1	−1	z R_z	
E	2	−1	0	$(x, y)(R_x, R_y)$	$(x^2 - y,^2\ xy)(xz, yz)$

4. The C_{nv} groups

C_{2v}	E	C_2	$\sigma_v(xz)$	$\sigma_v'(yz)$		
A_1	1	1	1	1	z	x^2, y^2, z^2
A_2	1	1	-1	-1	R_z	xy
B_1	1	-1	1	-1	x, R_y	xz
B_2	1	-1	-1	1	y, R_x	yz

C_{3v}	E	$2C_3$	$3\sigma_v$		
A_1	1	1	1	z	$x^2 + y^2, z^2$
A_2	1	1	-1	R_z	
E	2	-1	0	(x, y) (R_x, R_y)	$(x^2 - y^2\ xy)$ (yz, xy)

C_{4v}	E	$2C_4$	C_2	$2\sigma_v$	$2\sigma_d$		
A_1	1	1	1	1	1	z	$x^2 + y^2, z^2$
A_2	1	1	1	-1	-1	R_z	
B_1	1	-1	1	1	-1		$x^2 - y^2$
B_2	1	-1	1	-1	1		xy
E	2	0	-2	0	0	$(x, y)(R_x, R_y)$	(xz, yz)

C_{5v}	E	$2C_5$	$2C_5^2$	$5\sigma_v$		
A_1	1	1	1	1	z	$x^2 + y^2, z^2$
A_2	1	1	1	-1	R_z	
E_1	2	$2\cos 72°$	$2\cos 144°$	0	(x, y) (R_x, R_y)	(xz, yz)
E_2	2	$2\cos 144°$	$2\cos 72°$	0		$x^2 - y^2, xy$

C_{6v}	E	$2C_6$	$2C_3$	C_2	$3\sigma_v$	$3\sigma_d$		
A_1	1	1	1	1	1	1	z	$x^2 + y^2, z^2$
A_2	1	1	1	1	-1	-1	R_z	
B_1	1	-1	1	-1	1	-1		
B_2	1	-1	1	-1	-1	1		
E_1	2	1	-1	-2	0	0	(x, y) (R_x, R_y)	(xy, yz)
E_2	2	-1	-1	2	0	0		$x^2 - y^2, xy$

5. The C_{2h} group

C_{2h}	E	C_2	i	σ_h		
A_g	1	1	1	1	R_z	x^2, y^2, z^2, xy
B_g	1	−1	1	−1	R_x, R_y	(xz, yz)
A_u	1	1	−1	−1	z	
B_u	1	−1	−1	1	x, y	

6. The D_{nh} groups

D_{2h}	E	$C_2(z)$	$C_2(y)$	$C_2(x)$	i	$\sigma(xy)$	$\sigma(xz)$	$\sigma(yz)$		
A_g	1	1	1	1	1	1	1	1		x^2, y^2, z^2
B_{1g}	1	1	−1	−1	1	1	−1	−1	R_z	xy
B_{2g}	1	−1	1	−1	1	−1	1	−1	R_y	xz
B_{3g}	1	−1	−1	1	1	−1	−1	1	R_x	yz
A_u	1	1	1	1	−1	−1	−1	−1		
B_{1u}	1	1	−1	−1	−1	−1	1	1	z	
B_{2u}	1	−1	1	−1	−1	1	−1	1	y	
B_{3u}	1	−1	−1	1	−1	1	1	−1	x	

D_{3h}	E	$2C_3$	$3C_2$	σ_h	$2S_3$	$3\sigma_v$		
A_1'	1	1	1	1	1	1		$x^2 + y^2, z^2$
A_2'	1	1	−1	1	1	−1	R_z	
E'	2	−1	0	2	−1	0	(x, y)	$(x^2 - y^2, xy)$
A_1''	1	1	1	−1	−1	−1		
A_2''	1	1	−1	−1	−1	1	z	
E'	2	−1	0	−2	1	0	(R_x, R_y)	(xz, yz)

D_{4h}	E	$2C_4$	C_2	$2C_2'$	$2C_2''$	i	$2S_4$	σ_h	$2\sigma_v$	$2\sigma_d$		
A_{1g}	1	1	1	1	1	1	1	1	1	1		$x^2 + y^2, z^2$
A_{2g}	1	1	1	−1	−1	1	1	1	−1	−1	R_z	
B_{1g}	1	−1	1	1	−1	1	−1	1	1	−1		$x^2 - y^2$
B_{2g}	1	−1	1	−1	1	1	−1	1	−1	1		xy
E_g	2	0	−2	0	0	2	0	−2	0	0	(R_x, R_y)	(xz, yz)
A_{1u}	1	1	1	1	1	−1	−1	−1	−1	−1		
A_{2u}	1	1	1	−1	−1	−1	−1	−1	1	1	z	
B_{1u}	1	−1	1	1	−1	−1	1	−1	−1	1		
B_{2u}	1	−1	1	−1	1	−1	1	−1	1	−1		
E_u	2	0	−2	0	0	−2	0	2	0	0	(x, y)	

D_{5h}	E	$2C_5$	$2C_5^2$	$5C_2$	σ_h	$2S_5$	$2S_5^3$	$5\sigma_v$		
A_1'	1	1	1	1	1	1	1	1		x^2+y^2, z^2
A_2'	1	1	1	−1	1	1	1	−1	R_z	
E_1'	2	$2\cos 72°$	$2\cos 144°$	0	2	$2\cos 72°$	$2\cos 144°$	0	(x, y)	
E_2'	2	$2\cos 144°$	$2\cos 72°$	0	2	$2\cos 144°$	$2\cos 72°$	0		(x^2-y^2, xy)
A_1''	1	1	1	1	−1	−1	−1	−1		
A_2''	1	1	1	−1	−1	−1	−1	1	z	
E_1''	2	$2\cos 72°$	$2\cos 144°$	0	−2	$-2\cos 72°$	$-2\cos 144°$	0	(R_x, R_y)	(xy, yz)
E_2''	2	$2\cos 144°$	$2\cos 72°$	0	−2	$-2\cos 144°$	$-2\cos 72°$	0		

D_{6h}	E	$2C_6$	$2C_3$	C_2	$3C_2'$	$3C_2''$	i	$2S_3$	$2S_6$	σ_h	$3\sigma_d$	$3\sigma_v$		
A_{1g}	1	1	1	1	1	1	1	1	1	1	1	1		x^2+y^2, z^2
A_{2g}	1	1	1	1	−1	−1	1	1	1	1	−1	−1	R_z	
B_{1g}	1	−1	1	−1	1	−1	1	−1	1	−1	1	−1		
B_{2g}	1	−1	1	−1	−1	1	1	−1	1	−1	−1	1		
E_{1g}	2	1	−1	−2	0	0	2	1	−1	−2	0	0	(R_x, R_y)	(xz, yz)
E_{2g}	2	−1	−1	2	0	0	2	−1	−1	2	0	0		(x^2-y^2, xy)
A_{1u}	1	1	1	1	1	1	−1	−1	−1	−1	−1	−1		
A_{2u}	1	1	1	1	−1	−1	−1	−1	−1	−1	1	1	z	
B_{1u}	1	−1	1	−1	1	−1	−1	1	−1	1	−1	1		
B_{2u}	1	−1	1	−1	−1	1	−1	1	−1	1	1	−1		
E_{1u}	2	1	−1	−2	0	0	−2	−1	1	2	0	0	(x, y)	
E_{2u}	2	−1	−1	2	0	0	−2	1	1	−2	0	0		

7. The D_{nd} groups

D_{2d}	E	$2S_4$	C_2	$2C_2'$	$2\sigma_d$		
A_1	1	1	1	1	1		x^2+y^2, z^2
A_2	1	1	1	-1	-1	R_z	
B_1	1	-1	1	1	-1		x^2-y^2
B_2	1	-1	1	-1	1	z	xy
E	2	0	-2	0	0	$(x, y)\,(R_x, R_y)$	(xz, yz)

D_{3d}	E	$2C_3$	$3C_2$	i	$2S_6$	$3\sigma_d$		
A_{1g}	1	1	1	1	1	1		x^2+y^2, z^2
A_{2g}	1	1	-1	1	1	-1	R_z	
E_g	2	-1	0	2	-1	0	(R_x, R_y)	$(x^2-y^2, xy)(xz, yz)$
A_{1u}	1	1	1	-1	-1	-1		
A_{2u}	1	1	-1	-1	-1	1	z	
E_u	2	-1	0	-2	1	0	(x, y)	

D_{4d}	E	$2S_8$	$2C_4$	$2S_8^3$	C_2	$4C_2'$	$4\sigma_d$		
A_1	1	1	1	1	1	1	1		x^2+y^2, z^2
A_2	1	1	1	1	1	-1	-1	R_z	
B_1	1	-1	1	-1	1	1	-1		
B_2	1	-1	1	-1	1	-1	1	z	
E_1	2	$\sqrt{2}$	0	$-\sqrt{2}$	-2	0	0	(x, y)	
E_2	2	0	-2	0	2	0	0		(x^2-y^2, xy)
E_3	2	$-\sqrt{2}$	0	$\sqrt{2}$	-2	0	0	(R_x, R_y)	(xz, yz)

D_{5d}	E	$2C_5$	$2C_5^2$	$5C_2$	i	$2S_{10}^3$	$2S_{10}$	$5\sigma_d$		
A_{1g}	1	1	1	1	1	1	1	1		x^2+y^2, z^2
A_{2g}	1	1	1	-1	1	1	1	-1	R_z	
E_{1g}	2	$2\cos 72^0$	$2\cos 144^0$	0	2	$2\cos 72^0$	$2\cos 144^0$	0	(R_x, R_y)	(xz, yz)
E_{2g}	2	$2\cos 144^0$	$2\cos 72^0$	0	2	$2\cos 144^0$	$2\cos 72^0$	0		(x^2-y^2, xy)
A_{1u}	1	1	1	1	-1	-1	-1	-1		
A_{2u}	1	1	1	-1	-1	-1	-1	1	z	
E_{1u}	2	$2\cos 72^0$	$2\cos 144^0$	0	-2	$-2\cos 72^0$	$-2\cos 144^0$	0	(x, y)	
E_{2u}	2	$2\cos 144^0$	$2\cos 72^0$	0	-2	$-2\cos 144^0$	$-2\cos 72^0$	0		

8. The cubic groups

T_d	E	$8C_3$	$3C_2$	$6S_4$	$6\sigma_d$		
A_1	1	1	1	1	1		$x^2 + y^2 + z^2$
A_2	1	1	1	−1	−1		
E	2	−1	2	0	0		$(2z^2 - x^2 - y^2,$ $x^2 - y^2)$
T_1	3	0	−1	1	−1	(R_x, R_y, R_z)	
T_2	3	0	−1	−1	1	(x, y, z)	(xy, xz, yz)

O_h	E	$8C_3$	$6C_2$	$6C_4$	$2C_2(C_4{}^2)$	i	$6S_4$	$8S_6$	$3\sigma_h$	$6\sigma_d$		
A_{1g}	1	1	1	1	1	1	1	1	1	1		$x^2 + y^2 + z^2$
A_{2g}	1	1	−1	−1	1	1	−1	1	1	−1		
E_g	2	−1	0	0	2	2	0	−1	2	0		$(z^2, x^2 - y^2)$
T_{1g}	3	0	−1	1	−1	3	1	0	−1	−1	(R_x, R_y, R_z)	
T_{2g}	3	0	1	−1	−1	3	−1	0	−1	1		(xz, yz, zy)
A_{1u}	1	1	1	1	1	−1	−1	−1	−1	−1		
A_{2u}	1	1	−1	−1	1	−1	1	−1	−1	1		
E_u	2	−1	0	0	2	−2	0	1	−2	0		
T_{1u}	3	0	−1	1	−1	−3	−1	0	1	1	(x, y, z)	
T_{2u}	3	0	1	−1	−1	−3	1	0	1	−1		

Direct Product Tables

Direct Product Rules For Chemically Important Groups

1   General rules

χ	$'$	$''$
$'$	$'$	$''$
$''$		$'$

c	g	u
g	g	u
u		g

Unless otherwise indicated

χ	1	2
1	1	2
2		1

2   For C_2, D_3, C_{2v}, C_{3v}, C_{6v}, C_{2h}, D_{3h}, D_{6h}, D_{3d}

χ	A_1	A_2	B_1	B_2	E_1	E_2
A_1	A_1	A_2	B_1	B_2	E_1	E_2
A_2		A_1	A_1	A_2	E_1	E_2
B_1				A_1	E_2	E_1
B_2				A_1	E_2	E_1
E_1					$A_1 + E_2$	$B_1 + B_2 + E_1$
E_2						$A_1 + E_2$

3   For D_2, D_{2h}

χ	A	B_1	B_2	B_3
A	A	B_1	B_2	B_3
B_1		A	B_3	B_2
B_2			A	B_1
B_3				A

4   For C_{4v}, C_{4h}, D_{2d}

χ	A_1	A_2	B_1	B_2	E
A_1	A_1	A_2	B_1	B_2	E
A_2		A_1	B_2	B_1	E
B_1			A_1	A_2	E
B_2				A_1	E
E					$A_1 + B_1 + B_2$

5   For C_{5v}, D_{5h}, D_{5d}

χ	A_1	A_2	E_1	E_2
A_1	A_1	A_2	E_1	E_2
A_2		A_1	E_1	E_2
E_1			$A_1 + E_2$	$E_1 + E_2$
E_2				$A_1 + E_1$

6   For O_h, T_d

χ	A_1	A_2	E	T_1	T_2
A_1	A_1	A_2	E	T_1	T_2
A_2		A_1	E	T_2	T_1
E			$A_1 + E$	$T_1 + T_2$	$T_1 + T_2$
T_1				$A_1 + E + T_2$	$A_2 + E + T_1 + T_2$
T_2					$A_1 + E + T_2$